CHEMICAL ENGINEERING METHODS AND TECHNOLOGY

CHEMICAL VAPOR DEPOSITION (CVD)

METHODS AND TECHNOLOGIES

CHEMICAL ENGINEERING METHODS AND TECHNOLOGY

Additional books and e-books in this series can be found on Nova's website under the Series tab.

CHEMICAL ENGINEERING METHODS AND TECHNOLOGY

CHEMICAL VAPOR DEPOSITION (CVD)

METHODS AND TECHNOLOGIES

LEVI KARLSSON
EDITOR

nova
science publishers
New York

Library of Congress Cataloging-in-Publication Data

ISBN: 978-1-53619-949-9

Published by Nova Science Publishers, Inc. † New York

CONTENTS

PREFACE

Chemical vapor deposition (CVD) refers to a vacuum deposition method used to produce high quality, high-performance, solid materials in a variety of manufacturing industries. Chapter One provides a critical review of published experimental data for thin films of silicon nitride deposited by thermal and plasma CVD, plasma CVD, high density plasma CVD, atomic layer-by-layer deposition, and hot-wire CVD. Chapter Two describes several aspects of the use of CVD for single-crystal diamond synthesis for electronics. Chapter Three describes the properties of graphene and its preparation by a number of methods with a focus on the classical CVD method on copper foil together with graphene transfer onto a dielectric substrate.

Chapter 1 - This chapter provides a critical review of published experimental data for thin films of silicon nitride deposited by thermal and plasma chemical vapor deposition (CVD), plasma chemical vapor deposition (PECVD), high density plasma CVD (HDP-CVD), atomic layer-by-layer deposition (ALD), hot-wire chemical vapor deposition (HW-CVD). The data has been analyzed historically with a search depth of about 25 years. The focus is on the aspects of film growth. The preparation of thin films is described and the possibilities of their improvement are analyzed. The relationship between the composition of the film and physicochemical properties with the parameters of the deposition processes is considered. Comparative data on the properties of the films obtained using the above methods are presented. It is shown that the main properties of the films depend on the deposition temperature. The hydrogen content in the films can reach 40 at % for low-temperature PECVD. On the contrary, high deposition temperatures ensure the stoichiometry of the film with hydrogen content below ~ 5 at %. It is shown that the wet etching rate and film density are determined by the hydrogen content in the films. Performed review provides understanding of the differences of SiNH-contained thin film CVD methods in terms of the constant of the deposition rate, the effectiveness of reactant usage, and the film conformality on complicated reliefs of electronic devices.

Chapter 2 - The existing technologies of single-crystal diamond synthesis for electronics are expensive and have significant limitations. For example, artificial diamonds obtained at high pressure and high temperature (HPHT), require high energy consumption, have sectorial structure and metal catalyst inclusions. CVD-synthesis of such diamonds requires single crystal substrate and does not provide overgrowth with maintaining high quality of the diamond. At the same time, there is a factor which is known in the known diamond synthesis technologies, but is not explicitly taken into account in theoretical models. It is the temperature rise rate in space or on time. In the 2000s this factor (hereafter fast-heating criterion) was proposed as an empirical criterion able to propose new methods of the diamond synthesis, free of both high pressure and a previously prepared substrate.

In this chapter the authors try to trace the mechanism underlying this criterion, in particular case of the CVD diamond growth technologies, enhanced by activation of the gas flow (MWCVD, HFCVD), and to assess the prospects of the old idea at a new level.

Analysis of the fast heating shows, that the density of the nucleation source which is a vapour of small carbon species formed from methane-carbon-oxygen mixture, follows the temperature rise with a much smaller delay than the density of the same species as the saturated vapour components above the cluster surface.

Therefore, the synthesis of the carbon clusters of both graphite and diamond types occurs under conditions of supersaturation of the small hydrocarbon vapor, initially enormous but rapidly decreasing then. The supersaturation is much higher in the diamond channel than in the graphene (or graphite) one. In addition, in the presence of an oxidizing agent, oxidation of the diamond nuclei occurs much lower than that of the graphite nuclei. These results are associated with the fact that both oxidation and desorption of the carbon species from the surface of the diamond clusters are accompanied by breaking larger number of carbon bonds; however, the difference in the oxidation factor is much stronger than the difference in the desorption factor.

In the above consideration of the MWCVD process, the substrate is involved only as a factor which determine (in combination with the plasma) the temperature dynamics in space. The diamond nuclei arising in the gas phase are deposited on the substrate as already formed clusters. However, the fast-heating criterion plays an important role in the CVD diamond synthesis technology anyway. Therefore, considering the ideology of CVD synthesis of high-quality massive diamonds to be fundamentally exhausted, this criterion can be applied to other possibilities of obtaining diamonds. The idea of non-stationary temperature dynamics and ability to use it to control the channels of gas-phase condensation can be realised under the synthesis not on a substrate, but in the bulk of a medium, which excludes free diamond growth.

Such a medium may be a carbonate-silicate melt, in which the carbon cluster tends to develop as the diamond cluster rather than the graphite one. In this case experimental conditions are possible under which the fast-heating criterion is valid in the melt layer through which the gas mixture is passed.

The most significant of these conditions such as the necessary nucleation temperature of the diamond clusters and the temperature rise rate from the initial melting temperature to this nucleation temperature, are achievable under nonstationary HF induction heating at moderate parameters.

The values of these parameters practically do not differ from that of the MWCVD process, with the same simplifications that were used to describe the MWCVD. The nucleation temperature is about 1800-1900 K, but this value is based on a number of energy parameters obtained from quantum chemistry calculations and is estimated with an accuracy of ±400 K. So, the window of both the cluster nucleation and survival of the clusters is rather narrow and should be determined experimentally.

Chapter 3 - The presented text describes the properties of graphene, its preparation by a number of methods with the focus on the classical Chemical Vapour Deposition Method (CVD) on copper foil together with graphene transfer onto a dielectric substrate. It presents the results published by the author in a number of cited publications. It is thus an overview work, which can serve as a study material for those, who are beginning to get acquainted with graphene, its properties and preparation by the CVD method, which is frequently used in practice.

In: Chemical Vapor Deposition (CVD)
ISBN: 978-1-53619-949-9
Editor: Levi Karlsson

Chapter 1

CVD Technologies and Basic Properties of SiNH-Contained Thin Films for Applications in Electronic Devices

Vladislav Yu. Vasilyev*, PhD
Semiconductors and Microelectronics Department,
Faculty of Radio Engineering and Electronics,
Novosibirsk State Technical University, Novosibirsk, Russia

Abstract

This chapter provides a critical review of published experimental data for thin films of silicon nitride deposited by thermal and plasma chemical vapor deposition (CVD), plasma chemical vapor deposition (PECVD), high density plasma CVD (HDP-CVD), atomic layer-by-layer deposition (ALD), hot-wire chemical vapor deposition (HW-CVD). The data has been analyzed historically with a search depth of about 25 years. The focus is on the aspects of film growth. The preparation of thin films is described

* Corresponding Author's Email: vasilev@corp.nstu.ru.

and the possibilities of their improvement are analyzed. The relationship between the composition of the film and physicochemical properties with the parameters of the deposition processes is considered. Comparative data on the properties of the films obtained using the above methods are presented. It is shown that the main properties of the films depend on the deposition temperature. The hydrogen content in the films can reach 40 at % for low-temperature PECVD. On the contrary, high deposition temperatures ensure the stoichiometry of the film with hydrogen content below ~ 5 at %. It is shown that the wet etching rate and film density are determined by the hydrogen content in the films. Performed review provides understanding of the differences of SiNH-contained thin film CVD methods in terms of the constant of the deposition rate, the effectiveness of reactant usage, and the film conformality on complicated reliefs of electronic devices.

1. Introduction

Thin-film inorganic materials are the base for electronic devices manufactured using integrated microelectronics technologies, nano- and microsystem technology. Chronologically, thin films (TF) of silicon dioxide and silicate glasses, silicon nitride, and polycrystalline silicon began firstly to be used in the technology of integrated circuits (IC) [1]. The chemical vapor deposition from the gas phase (CVD) takes the leading place in the creation of TF. Using CVD, the necessary inorganic material is synthesized by chemical reactions directly during the multistage technological cycles of IC device manufacturing. The sum of general issues of CVD of thin films about the IC technology is detailed in [2]. In [3], the main results, trends, and issues of the development of CVD methods of thin films for advanced IC and nano- and microelectromechanical systems (MEMS) are outlined.

Silicon nitride thin films (SNTF) usually assumed to correspond to the stoichiometric formula Si_3N_4 (Si:N = 1:1.33) are one of the most popular dielectrics in the IC and MEMS technology. The use of SNTF is based on the presence in Si_3N_4 of the several unique properties, for example, resistance to aggressive chemical acids, such as hydrofluoric (HF), or hot orthophosphoric (H_3PO_4), differences in plasma etch rates in fluorine-containing gas mixtures from those for silicon dioxide, which ensures selectivity of etching, increased hardness, gettering properties for sodium ions undesirable for IC technology, etc. [4, 5]. The methods and equipment for the production of SNTF at CVD by ammonolysis reactions of silicon compounds (reactions with gaseous ammonia, NH_3) have been implemented for many years in the mass production of the IC and MEMS. However, the development of the integrated technologies poses new tasks to optimization and modification of the existing processes / compositions / properties of SNTF and development of new methods for CVD of the SNTF. CVD technology and, especially, TF deposition temperature, directly defines the film composition, structure, and properties.

In the case of the SNTF production by CVD methods, the use of the stoichiometric formula Si_3N_4 is not unambiguous. The composition of the resulting thin film material depends on the initial chemical reagents and the conditions of CVD, the main one being temperature. For example, elements from the initial chemical reagents (chlorine, oxygen, carbon) can present in the SNTF as residues. However, besides silicon and nitrogen, the most common component of SNTF is hydrogen, whose source is the starting reagents. The concentration of hydrogen in SNTF can be from several units [5-7] (high CVD temperatures) to dozens of atomic percent (low CVD temperatures) [5]. This is important for many applications, including for MEMS [6, 7]. Thus, to characterize the actual composition of SNTF in a general form, it is more correct to use the general formula $Si_xN_yH_z$. In the literature, such notations for films as SiN_x [8] and α-SiN_x [9] for films with a hydrogen concentration less than 5 at % (below these SNTFs are designated as stoichiometric, S-SNTFs and the term SNTF will be used as a more general) and α-SiN_x:H [9] and α-SiN:H [10] for non-stoichiometric films with a hydrogen concentration greater than 5 at % are encountered (below the author uses the term “silicon-nitrogen-hydrogen-containing thin films, SiNH-TFs”).

At high-temperature annealing, the concentration of hydrogen in SNTF decreases. Understanding film composition can be indirectly, but quickly, made by the value of the refractive index (RI). For stoichiometric Si_3N_4, the RI is 2.02 ± 0.02; the values above correspond to the enrichment of material with the silicon, the values below correspond to the enrichment of the material, as usually, with the oxygen [4, 5].

The purpose of this chapter is to analyze the status and trends in the development of technologies for the production of SNTF based on information sources for the last three decades. The information base for the review is very large. The main attention is paid to SNTF deposition, composition, and basic physical and chemical properties. In the sections below, the author reviews separately different types of CVD SNTF in a sequence of deposition temperature decreasing: from high-temperature thermally-activated to low-temperature plasma- and catalytic-activated CVD. After this consideration, basic film properties are summarized and discussed.

2. Thermally-Activated CVD in Flow-Type Reactors

2.1. Brief Description of Thin Film CVD Processes

Typically, CVD processes are used in flow type reactors with simultaneous and continuous reactant injection and continuous removal of reaction by-products. These processes are usually characterized by the following basic set of CVD parameters: deposition temperature (T_d, °C), deposition pressure (P_d, Torr), deposition time (t_d, min or s), as well as partial pressures of reactants, continuous flow rates of reactants and total gas flow rate, etc. The CVD processes are carried out with a simultaneous reactant injection into the reaction chambers and total constant flow rate at selected process pressure P_d, as shown schematically in Figure 1(a). Therefore, the CVD film thickness (d) obeys the equation $d = W_d \times t_d$, where the CVD deposition rate (W_d) is usually constant during the process. The dependence of the deposition rate on the parameters of the CVD process (for example, reactants A and B with concentrations [A] and [B]) corresponds to the classical chemical equation $W_d = k[A]^n[B]^m$. Here k is the reaction constant with an exponential dependence on temperature ($k = k_0 \times exp(-E_a/RT_d)$ and E_a is the activations energy, R is the gas constant. The coefficients n and m reflect the order of the chemical reaction for reactants A and B. The accumulated knowledge about studied silicon-based films shows that most CVD reactions exhibit close to first-order reactions for silicon compounds (precursors), i.e., $n \sim 1$.

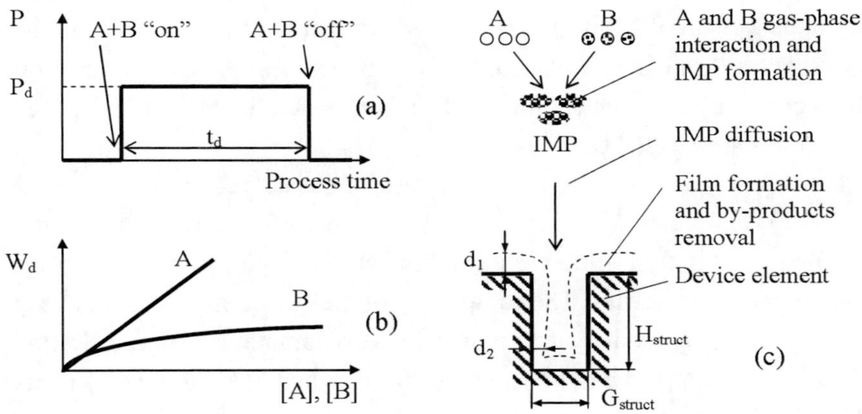

Figure 1. The main characteristics of thin film CVD processes with reactants A and B: constant P_d along with deposition time t_d providing constant W_d value (a); concentration dependences of W_d (b); a simplified scheme of the CVD process with the formation of IMPs in the gas phase and their diffusion with subsequent consumption on the surface with the formation of more or less non-conformal thin film depending on the ratio of the rates of IMP formation, diffusion, consumption, and AR values (c).

In contrast, optimized thin film CVD processes are performed in a large excess of the second reaction components for ensuring the stoichiometry of the reaction products. This correlates in most cases with the "pseudo-zero" order of the chemical reaction with the second reaction component, Figure 1(b). The latter simplifies the above equation as follows: $W_d \sim k_{eff} \times [C_{Si}]$, where $[C_{Si}]$ is silicon-contained reactant concentration, mol/l. In this equation, k_{eff} reflects the coefficient named "effective constant of the deposition rate" with the dimension of the first-order reaction kinetics cm/s. Its values are in the range of 0.1 to 10 cm/s for studied silicon-based thin films [2, 3]. The lowest k_{eff} values were found for CVD processes described by heterogeneous reaction mechanisms such as the oxidation of TEOS in

LPCVD reactor. The highest k_{eff} values were found for high deposition rates CVD processes such as chain-branched monosilane-oxygen APCVD and LPCVD processes, and especially for PECVD processes and unbiased HDP-CVD processes. Details regarding k_{eff} rating can be found elsewhere [2].

Schemes of CVD processes are described with consecutive stages of formation of intermediate products (IMP) from simultaneously introduced reactants followed by diffusion of IMP to the substrate surface and continuous growth of a thin film, see the scheme in Figure 1(c). It should be noted that the so-called "pulsed CVD" processes, where one reaction component can be introduced in short pulses, generally correspond to the above description of CVD.

The CVD processes listed above met all basic requirements of IC technology until the late 1990s, including high deposition rates and throughput, good thin film conformality, and the gap-fill capability in existent device geometries. The complexity of the device structure geometry is usually characterized by the dimensionless parameter "Aspect Ratio" (AR_{struct}), which represents the ratio of the gap height (H_{struct}, mm) to its width (G_{struct}, mm), Figure 1(c). Usually, simple single device steps or gaps in IC device elements are used to determine the thin film conformality (or step coverage). In the most common case, it is defined as the ratio of the film thickness at the bottom of a device step (or gap) to the film thickness at the top of this step (or gap), d_2/d_1 (%), Figure 1(c). The quantitative description of the CVD process made it possible to relate the experimentally obtained value of the film step coverage with the k_{eff} value for this CVD process, regardless of the reactors used. It was shown that, depending on the CVD process and its conditions, film conformality ranges from a few percent (for high-speed and plasma-activated CVD processes) to almost 100% for relatively slow heterogeneous CVD processes [2, 3]. Gap filling is a characteristic of a device structure that can be filled with CVD films without any unwanted voids (emptiness) inside. The reason for the formation of voids and their sizes is a consequence of the non-conformal growth of thin films. The conformality and gap-fill characteristics of thin film CVD processes are evaluated using scanning electron microscopy (SEM) cross-sectional images of device structures coated with deposited films.

2.2. Industrial Technologies for Production of S-SNTF by CVD

S-SNTF processes include processes at 750–900°C in “batch” (“group”) low-pressure tubular flow reactors with “hot walls,” Figure 2(a,b).

The processes (well-known as “Low Pressure CVD, LPCVD”) are realized due to the thermal activation of chemical reactions of ammonolysis of monosilane or its chlorinated derivatives. The hydrogen concentration in such SNTF is usually less than 5 at % [5–7] and they can be considered as stoichiometric. Such S-SNTFs are required, for example, for use in the sub-route of IC manufacturing, known as Front-End-of-Line (FEOL), and allowing high processing temperatures. In the course of FEOL, the transistor IC structures are formed with elements of their isolation using high-temperature oxidation of sites in silicon (Local Oxidation of Silicon, LOCOS), insulating trenches in silicon with their filling with silicon dioxide (Shallow Trench Isolation, STI), insulating regions (spacer) for the transistor gates. In contrast, the sub-route of IC manufacturing, known as the “Back-End-Of-Line” (BEOL) – formation of a multilevel system for metallization of ICs, requires only low temperatures of technological operations, including a deposition. In MEMS technologies, S-SNTFs are used also as protective coatings on silicon, if long-term etching of open silicon regions by anisotropic etching in aqueous KOH solutions is necessary.

The gross reactions of ammonolysis of silicon compounds and the main characteristics of the corresponding CVD processes for the production of S-SNTF are given below. The reaction of ammonolysis of monosilane (SiH_4, gas under normal conditions) is described by the scheme (1):

$$3\ SiH_4 + 4\ NH_3 \rightarrow Si_3N_4 + 12\ H_2 \quad (1)$$

The advantage of the process, realized in the range 750 – 800°C, pressures < 130 Pa, the ratio NH_3/SiH_4 > ~50, is the lack of accumulation of reaction by-products on the walls of the reactor and the vacuum system. Disadvantages of the process:

- low productivity: the so-called flow-diffusion hot-wall LPCVD reactor is needed to increase it, which is a cylindrical tube, inside which a perforated cylindrical cage with substrates is placed [2], see Figure 1(c);
- non-uniformity of a thickness of S-SNTF on substrates;
- non-conformality of growth on the relief of IC;
- macro-defects in S-SNTFs.

The reaction of ammonolysis of tetrachlorosilane ($SiCl_4$, a liquid under normal conditions) is described by the scheme (2):

$$3\ SiCl_4 + 16\ NH_3 \rightarrow Si_3N_4 + 12\ NH_4Cl \qquad (2)$$

The advantages of the process carried out in the LCVD at 800 – 900°C, the pressure of up to several hundred pascals, $NH_3/SiCl_4 > \sim 10$: high productivity (100 substrates per process) in a simple cylindrical LPCVD reactor, good homogeneity of S-SNTF and growth conformality on the ICs relief, absence of the macro-defects, the possibility of growth of a film with equal and uniform thickness at the same time from both sides of the substrates.

Disadvantages are:

- need to use a gas carrier to supply $SiCl_4$ vapor (nitrogen, argon);
- frequent need for chemical treatment (cleaning) of the reactor and the walls of the vacuum system to remove the precipitation like NH_4Cl.

The growth scheme of S-SNTF for the reaction of dichlorosilane ammonolysis (SiH_2Cl_2, gas under normal conditions) is displayed as follows (3):

$$3\ SiH_2Cl_2 + 10\ NH_3 \rightarrow Si_3N_4 + 6\ NH_4Cl + 6\ H_2 \qquad (3)$$

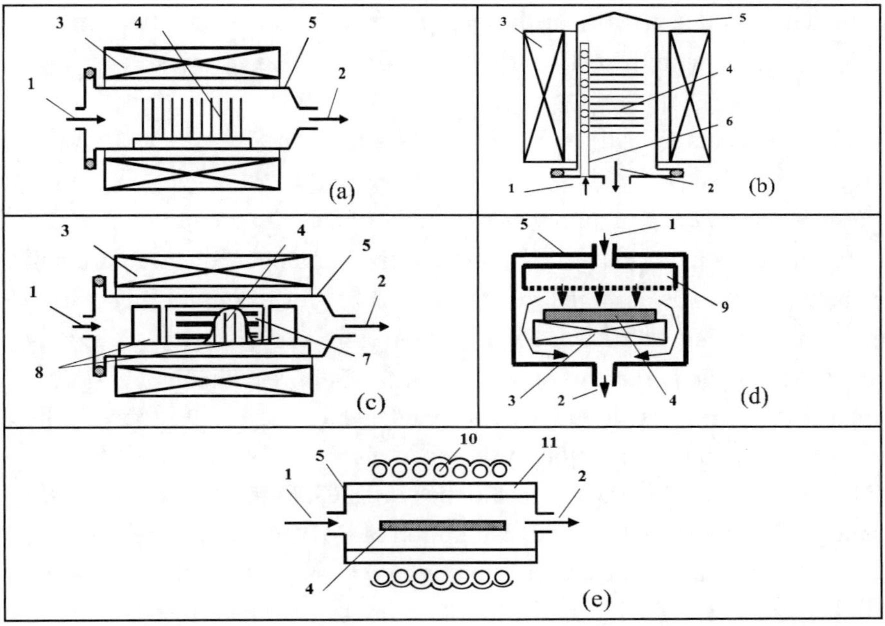

Figure 2. Simplified sketches of production flow reactors for stoichiometric silicon nitride deposition in hot-wall tubular CVD system with horizontal (a) and vertical (b) reactor orientation. Definitions: 1 – reactants "in"; 2 – reaction by-products "out"; 3 – heater; 4 – semiconductor wafer; 5 – reactor; 6 - distributed reactant feed (optional); 7 – substrate cage; 8 – cowls for gas-dynamic formation in a reactor; 9 – gas mixture assembly named "showerhead"; 10 – tungsten-halogen lamps; 11 – quartz window.

Ammonolysis of SiH_2Cl_2 is a standard industrial method for obtaining S-SNTF for IC and MEMS technologies implemented in cylindrical LPCVD reactor at 700 – 800°C, the pressure of about 50 Pa, $NH_3/SiH_2Cl_2 > \sim 5$. Advantages are generally similar to $SiCl_4$, but the amount of undesirable by-products is twice as little.

The considered methods of S-SNTF preparation have a good theoretical description. Peculiarities of the growth kinetics of S-SNTF in CVD with the participation of SiH_4, $SiCl_4$, and SiH_2Cl_2 are systematized and quantitatively described in [2]. The CVD processes are realized in the kinetic region, which is characterized by a linear dependence in the "Arrhenius coordinates" (logarithm of the reaction rate constant versus the reciprocal absolute temperature). In light of the scheme presented in Figure 1(c), a scheme of a universal two-stage sequential reaction with the stages of formation of an intermediate product (IMP) in the gas phase and its consumption on the substrate surface with S-SNTF formation is proposed as follows:

$$\text{Initial reagents} \rightarrow \text{IMP} \rightarrow \text{S-SNTF} + \text{by-products} \quad (4)$$

According to the ammonolysis reaction of SiH_4, the growth of S-SNTF is described as being limited by the stage of IMP homogeneous formation in the gas phase (it was assumed that this could be, for example, SiH_2, SiH_3, nitrogen-containing IMP, etc. [5]). In this case, the diffusion of slowly formed IMP to the surface of substrates in the bulk reactor, Figure 1(a), is critical and determines the above disadvantages of the CVD process. The CVD processes by the ammonolysis reactions of SiH_2Cl_2 and $SiCl_4$ with practically identical kinetic characteristics in LPCVD reactor are described as being limited by the stage of consumption of IMP on the substrate surface (heterogeneous stage). As the IMPs in this case there are the compounds like SiH_2ClNH_2 and $SiCl_3NH_2$, rapidly produced in the gas phase by the addition of ammonia to silicon tetrachloride with the elimination of HCl. As a result, a constant concentration of IMP is established in the CVD reactor, and the consequence of the CVD process limiting by heterogeneous stage results in the independence from diffusion processes. As a consequence, the process proceeds with small (about 1%) non-uniformity of the thickness and properties of S-SNTF films in the LPCVD reactor, and also with a very high (> 90%) conformance of their application to complex IC reliefs. The quantitative description of the S-SNTF growth patterns in the LPCVD reactor using the kinetics of successive irreversible first-order reactions in so-called “closed systems” [11] allowed us to adequately explain the set of observed similarities and differences in S-SNTF growth and suggest approaches for process modernization.

Comparison of severe differences in S-SNTF CVD kinetics is shown in Figure 3 [2]. This Figure allows clarifying the reaction mechanism experimentally. Comparative analysis of S-SNTF deposition rate profiles obtained for the ammonolysis of SiH_4, SiH_2Cl_2, and $SiCl_4$ at the same experimental conditions. Data in Figure 3 show significant differences between the kinetics of the ammonolysis of monosilane and its chlorinated derivatives. This figure also indicates the similar mechanisms of ammonolysis of the chlorides. The reactions of ammonolysis of SiH_2Cl_2 and $SiCl_4$ were characterized by close enough the effective rate constants values in the $W_{d(max)}$ regions (0.74 and 0.46 cm/s, respectively) and close conversion values of initial silicon reactants (26.6% and 18%, respectively). Both these characteristics are much lower than those for the ammonolysis of SiH_4 (4.9 cm/s and 46.5%).

It can be seen that "batch" LPCVD S-SNTF with the use of monosilane causes a lot of disadvantages, such as the film non-uniformity on in batch. Therefore, a temperature gradient is necessary to compensate for the deposition rate decrease along with the tube. In contrast, LPCVD using chlorides does not have such a problem and allows obtaining good film uniformity.

2.3. Directions for Improvement of the Technologies of SNTF by CVD

For flow reactors, four directions of the development of thermally activated processes of CVD SNTF were identified [12].

2.3.1. Improvement of the Existing Industrial Processes for S-SNTF in LPCVD Reactors

Typically, the industrial LPCVD process (mainly the process of dichlorosilane ammonolysis), is performed according to the base recipes. The improvements involve the optimization of the basic recipes, the most often – by adjusting temperature, pressure, and gas flow values [13, 14]. The aim is to improve the characteristics of the process for the production of S-SNTF, for example, to reduce the non-uniformity of the film thickness on the substrates, to increase the repeatability of the process, etc. In this case, the composition and properties of the material remain unchanged. Improving the conformality of TF deposition on structures with increased aspect ratios for heterogeneous processes can be optimized using the approach called the "Thiele criterion (modulus)" (taken from the theory of heterogeneous catalysis, relates diffusion processes and the rate constant of a heterogeneous reaction). This optimization can be achieved by reducing of pressure and temperature of the CVD processes [2]. There are also examples of modeling of these processes [15, 16].

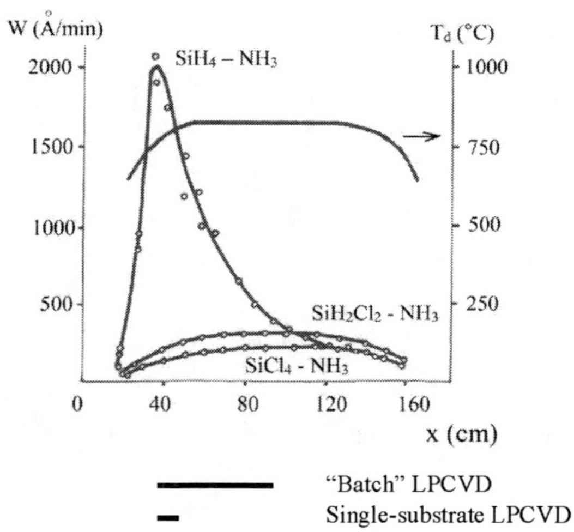

Figure 3. Comparative S-SNTF deposition rate profiles within a hot-wall tubular LPCVD reactor in $SiH_4 – NH_3$, $SiH_2Cl_2 – NH_3$ and $SiCl_4 – NH_3$ systems at $T_d = 825$ °C, $P_d = 332$ Pa, $\upsilon = 650$ cm/s, total gas flow 190 l/h; silicon compound concentrations 1.55×10^{-6} mol/l, ammonia concentration 2.66×10^{-5} mol/l, ammonia to silicon compound ratio – 17. These curves were obtained using a flat quartz boat with horizontal silicon wafer placement along with the gas flow. Equivalent tentative wafer placement for vertical "batch" LPCVD processes and single-substrate LPCVD reactor are shown as lines below the graph.

2.3.2. Modernization of the Processes for Obtaining Films in Flow CVD Reactors

Under the modernization (as a rule, for the base recipe), the principal changes of the parameters of CVD are meant, but without changing the initial reagents or the method of feeding the reagents in the CVD reactor. This is normal practice to search for new compositions and material properties. For MEMS, such property is mechanical stress in the films [17, 18]. In [17], silicon-enriched films with reduced values of mechanical stress were obtained in LPCVD under other invariable conditions 800 – 850°C and

30 – 70 Pa) by changing the SiH_2Cl_2/NH_3 ratio to 14 (with strong enrichment of the gas mixture with dichlorosilane). Further, such films are referred to as "silicon-enriched SNTF" or SE-SNTF. As a result of the increase in the SiH_2Cl_2/NH_3 ratio in the abovementioned limits, the refractive index of the films varied from a typical value for S-SNTF RI ~ 2.0 to silicon-enriched RI ~ 2.7. At the same time, the mechanical stress in SE-SNTF decreased steadily from 1.2 – 1.4 GPa to ~ – 0.15 GPa. Similar values of mechanical stress near 0.13 GPa are presented in [18], where the investigations were carried out at 835 – 875°C, 39 Pa pressure and a fixed value of gas flow ratio $SiH_2Cl_2/NH_3 = 10$. The wet etch rates for SE-SNTF in standard IC solutions: Buffered Oxide Etch, BOE (25°C), 40% HF (25°C), and H_3PO_4 (160°C) were about 2.5 – 3, and KOH (50°C) - about 5 times smaller than those for S-SNTF. Similar studies were performed in [19] for monosilane and ammonia in LPCVD reactor in the range of 725 – 775°C and the ratio $NH_3/SiH_4 = 0.2 – 1.6$. The changes in the mechanical stress in the films were from –0.5 GPa to +0.9 GPa.

The authors of these studies noted that when SE-SNTFs were obtained, the uniformity of the films deteriorated strongly along the length of the tubular LPCVD reactor. In the opinion of the author of the review, the reason is that low ratios of ammonia to the silicon-containing starting reagent and high temperatures of the CVD create conditions for the parallel development of the pyrolysis (thermal decomposition) reactions of the initial silicon-containing reagents SiH_4 and SiH_2Cl_2. The possibility of thermal decomposition of these reagents under similar experimental conditions of the components was investigated in [20, 21]. Simultaneously, the pyrolysis reactions that occur cause an accumulation of by-product concentrations along the length of the LPCVD reactor reaction zone. Thus, when SE-SNTF is deposited in the LPCVD reactor, the concentrations of the initial reagents and by-products will differ in each conventional section of the reactor working area. Accordingly, along the length of the reactors, the composition and properties of SE-SNTF will be different, indicating the inapplicability of such processes and reactors for the production of MEMS.

The author believes, that the plasma deposition at a temperature of about 400°C could serve as the best one for the preparation of SE-SNTF. Such possibility, in particular, the results of [22] indicate, where the authors investigated the synthesis and properties of semi-insulating polycrystalline silicon films obtained in an LPCVD reactor with an electrode system (details in [23]) with the ratio $NH_3/SiH_4 = 0.5 – 5$. Mechanical stresses of such films close to SE-SNTF were in the compression range in the band from – 0.9 to – 0.3 GPa. The advantage of this process is the absence of any thermally activated ammonolysis and thermal decomposition reactions at low deposition temperatures, i.e., the patterns of TF growth are mainly determined by the design of the equipment and the characteristics of the plasma discharge. Moreover, the production of films in individual plasma reactors (see below) can give even more uniform coatings, although the plasma processes have limitations on the conformality of deposition.

2.3.3. New Reactants for Production of S-SNTF in Flow CVD Reactors

The general aim of this type of research was to search for reactants that significantly reduce the deposition temperature. The authors [24] studied SNTF obtained in a "hot-wall" LPCVD reactor of tris(dimethylamino)silane $((CH_3)_2N)_3SiH$ and ammonia at a temperature of 650 – 900°C and pressure less than 130 Pa. The authors sought the so-called "environment friendly" - a silicon-containing reagent capable of substituting dangerous (SiH_4) and corrosive (SiH_2Cl_2, $SiCl_4$) reagents. The growth rate of films in the range of 650 – 800°C and pressure 20 – 80 Pa with the ratio $NH_3/((CH_3)_2N)_3SiH$ from 1 to 10 was 0.3 to 9.0 nm/min. The authors calculated that the optimum deposition temperature should be 800°C and the ratio $NH_3/((CH_3)_2N)_3SiH$ = 10. In the films, a carbon with a concentration of up to 5 at % and hydrogen (concentration was not shown) were detected.

The conformity of coatings on the IC device steps was evaluated for a structure with a depth of 1 μm and a width of 0.75 μm at a level of only 70%. In [25], the possibility of growth of SiN_x:H films at 250 – 500°C was studied with the participation of a mixture of the oligomer Si_2Cl_6 and hydrazine N_2H_4. The deposition rate was about 4 nm/min, and the refractive index was 1.67 – 1.75. The deposition rate was sharply reduced at temperatures below 350°C, which was explained by the formation of NH_4Cl-like materials. The concentration of hydrogen was estimated at 30 at %. The authors of [26] investigated the growth of TF from SiI_4 and ammonia in an atmospheric reactor at 350 – 550°C. The growth rates were about 80 nm/min, iodine concentration in the films was 0.8 – 4.2 at %, hydrogen concentration ~ 30 at %. The conformality of TF was estimated by the authors better than 85% on a structure with a size of 1.6 μm per 1 μm, however, with inclined walls. This makes it impossible to make adequate comparative estimates. In [27], the growth of films from hexamethyldisiloxane $(CH_3)_3SiOSi(CH_3)_3$ and ammonia at atmospheric pressure and 750 – 850°C was studied. At optimum deposition temperature of 800°C, the achieved refractive index of 1.88 conformed more to silicon oxynitride $Si_xO_yN_z$.

2.3.4. Single Substrate Flow Reactors for Production of S-SNTF by CVD

Two approaches need to be mentioned here, both directions were realized by Applied Materials, Inc., in equipment and CVD processes. Both directions assumed the use of monosilane ammonolysis reaction. Here need to highlight that CVD reaction conditions for single chambers differ totally, as shown in Figure 3. Equivalent substrate placement in a single chamber can be localized close to the reaction maximum $W_{d(max)}$, therefore providing good film uniformity with a very high deposition rate.

Simplified sketch of single the first type chamber is presented in Figure 1(d) [28]. A substrate is placed on the resistive heater, heated up to about 800°C, and in further exposed to a mixture of reagents SiH_4, NH_3 at significantly higher (13.3 – 66.5 GPa) pressure as compared to that in "batch" LPCVD reactors (less than 100 Pa) [29]. Deposition rates in this type of single changer were about 3 nm/s and CVD process duration is a few tens of seconds.

The second approach of CVD processes and single chamber are different. This approach is known as "Rapid thermal chemical vapor deposition, RTCVD." One of the principal advantages of RTCVD is a much lower thermal budget than furnace LPCVD processes.

A typical RTCVD system is shown in Figure 1(e) [30]. One of the key advantages of RTCVD is that a cold wall system is used which generates fewer particulates. A second advantage is that higher deposition rates are observed and, because the substrate is heated to a higher temperature, improvements in film quality are achieved. RTCVD also reduces the problems of autodoping and interdiffusion through a lower thermal budget than conventional LPCVD. RTCVD silicon nitride thin film characterization was done in [31]. In particular, the low hydrogen content of about 5 at % and good film stoichiometry were confirmed.

It is necessary to note that single substrate chambers provide generally one very important advantage: they allow adjust the film composition (for example, purposeful silicon enrichment) with uniform distribution within wafers that is not possible in batch "hot-wall" reactors even with distributed feed.

2.4. Conclusion to Section 2

Continuous improvement and development of IC and MEMS technologies pose challenges for thin film coating technologies. The main of them include the production of films of a given composition and properties, as well as the conformality of film growth on the high-aspect structures of modern devices. The review made it possible to single out the following trends in the development of technologies for the production of SNTF during the thermal activation of reactants in flow reactors. These include: search for new initial silicon-containing reactants, decrease in the temperature of the CVD processes, and search for new designs of CVD reactors. The solutions discussed in the literature for modifying the composition of thin films to obtain new properties required for MEMS applications require careful analysis of the possibilities of their hardware implementation to ensure homogeneity and reproducibility of the composition and properties of coatings on semiconductor substrates. At present, the investigated directions of CVD development in flow reactors have not found any prospects for a principal improvement in the conformality of thin films for complex 3D structures.

3. Thermally-Activated CVD in Cycle-Type Reactors

3.1. Brief General Description of Thin Film Cycle-Type Processes

In the early 2000s, continued shrinkage of the IC device elements and gaps between them reached a value of 100 nm. The CVD processes developed were found to exhibit some conformality and gap-fill worsening along with shrinkage of device structures. Therefore, it was expected that they would not be applicable to address the upcoming technology challenges. For solving the tasks of creation of the conformal coatings on the high-aspect ratio structures of modern IC and MEMS, it appears to be promising to use a version of the cyclic organization of the processes. In such process organization, the reagents are supplied to the typical reactors for CVD as the consecutive dosed pulses. In a general case, such CVD processes, for example, are named "pulsed-CVD" (P-CVD). It should be noted that P-CVD processes showed some advantages for short TF deposition processes due to precise control of pulse numbers. In this case, the total film thickness is a sum of thicknesses deposited in every process pulse. Note that the case when one reaction component is introduced in short pulses generally corresponds to the above description of CVD, see scheme in Figure 1(c).

Atomic Layer Deposition (ALD) technique [32] was shown to be promising due to completely different process implementation and features. The ALD technique is considered by some authors as a part of CVD (namely, P-CVD) with some peculiarities. Both techniques methods use very similar equipment (see Figure 2) and follow common thin film methodology. However, unlike the traditional CVD processes mentioned above and P-CVD, ALD processes are organized cyclically with a sequential injection of the precursor and other reactants separated by an inert gas, see ALD process cycle scheme in Figure 4.

This represents the thermally activated ALD (TA-ALD) process. In the case of the plasma-enhancement of the second reactant (PE-ALD), the plasma discharge is usually switched on some short time after the start of this reactant pulse and switched off at the end of the pulse. It can be seen that such cyclic organization of the process completely excludes any reactant interaction in the gas phase and the formation and diffusion of IMPs. All processes are localized on the surfaces of objects, even if they have very complex stepped surfaces. A key advantage of the ALD method was postulated as the accurate layer-by-layer deposition in the so-called "ALD-window" temperature region due to the self-limited chemical reaction (range with the near-zero temperature dependence of the deposition rate) and, therefore, excellent thin film conformality. Thus, ALD, in comparison with CVD, is a strongly surface-limited heterogeneous process. It should be noted that the evaluation of ALD processes in terms of k_{eff} is also possible using similar to CVD assumptions, such as deposition rates, reactant doses, reactor volume, etc. Such an assessment made by the author allowed estimation of an order of k_{eff} for ALD processes in the range of about 10^{-4} – 10^{-5} cm/s. These very small values explain the key ALD disadvantage, such as extremely low growth rates, not exceeding, in theory, a monolayer value per process cycle (but more often even as low as 0.01 nm per process cycle). In addition, the process pulse duration can sometimes be quite lengthy.

A schematic temperature dependence diagram of a growth per cycle (GPC or W_d) for ALD processes is shown in Figure 5(a) [33]. The following processes/regions are numbered: an increase of the deposition rate 1; a condensation region 2; desorption region 4; CVD region 5. Region 3 of the "ALD-window" for deposition processes is located between T_2 and T_1 and is characterized as close to a flat temperature dependence due to self-limited adsorption. Thus, according to this diagram, the process should be characterized as ALD if it is realized in the temperature range of the "ALD-window" ($\Delta T=T_2-T_1$). In this region, both reactants should exhibit GPC saturation relative to the reactant dose, Figure 5(b), which allows the determination of the appropriate reactant exposure values for complete self-limited adsorption of reactants. The required film thickness is achieved by increasing the number of deposition cycles, Figure 5(c). Typically, the GPC is constant during the processing time, but in some cases, delays in film growth (incubation) may occur in the early stages of the process.

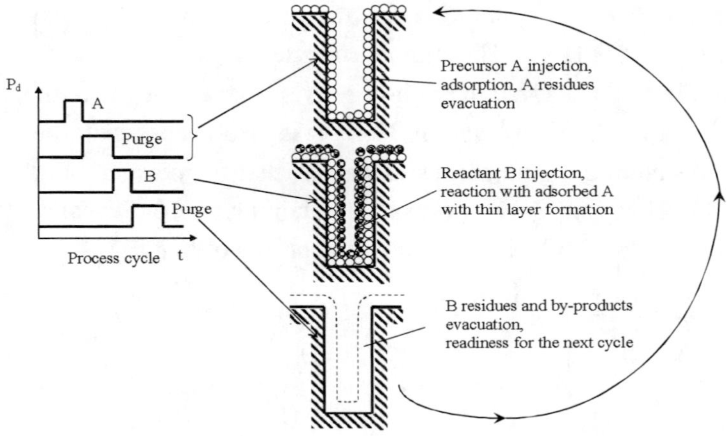

Figure 4. Diagram of a cyclic process with precursor A and reactant B and visualization of ALD processes on the surface of a stepped object.

Despite the very low deposition rate values compared to CVD processes, ALD processes are rightly considered a promising solution for device structures with an aggressive AR_{struct} in the era of micro- and nanotechnology [34, 35] including traditional and new dielectric materials [36] (SiN_x, SiO_2, SiC, C-containing SiN_x, and SiO_2).

Development of the cyclic processes for obtaining SNTF (just like for the CVD processes in the flow reactors goes in two directions:

Direction 1 includes research of the possibilities for realization of the thermally-activated industrial ALD processes at a lowered temperatures as for the industrial processes of CVD S-SNTF. In this case of interest is the possibility of the use of both the traditional industrial reagents and the new reagents. Exactly this direction is of great interest for the promising developments of IC and MEMS with the high aspect ratio structures.

Direction 2 includes the low-temperature methods for obtaining TF with an application, as a rule, of various versions of additional plasma activation of the nitrogen-containing reagents. The development of this direction is determined by new, mainly low-temperature directions for the application of TF. A considerable by volume [8] publication appeared where the authors attempted to cover a large array of data concerning mainly the plasma-activated SiNH-TF. The authors considered the films received in the traditional plasma reactors, in high-density plasma, and also the ALD methods with the plasma activation. Here, it is also necessary to underline that, notwithstanding the numerous assertions that the plasma methods for obtaining TF allow us to solve the tasks of obtaining the conformal coatings, so far, there are no convincing experimental proofs of that [37].

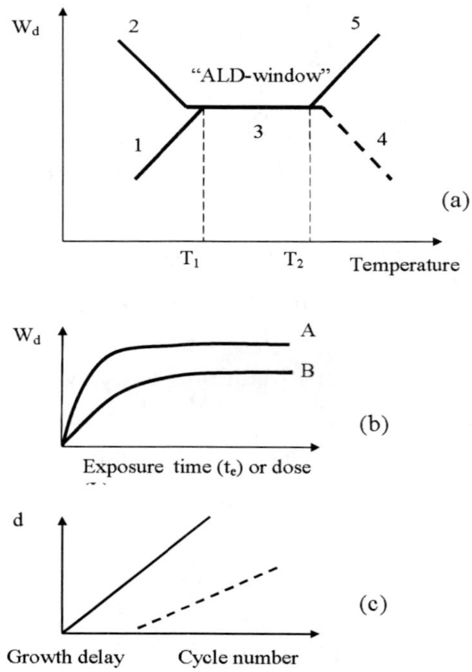

Figure 5. Conventional schematic of the dependence of the deposition rate on T_d (a), exemplary saturation dependences of GPC on precursor and reactant exposures time (t_e) or dose (L) (b), typical dependences of the thickness on the cycle number with and without delayed film growth (c).

In this section, we consider the main published results for thermally-activated SNTF according to Direction 1. Note that there have been some general reviews published on this topic [8, 9, 36, 38, 39].

3.2. Cyclic Deposition with the Use of the Traditional Industrial Reagents

As shown in section 2, the industrial processes for obtaining S-SNTF for IC and MEMS are realized in the flow LPCVD in the ammonolysis reactions of SiH_4, $SiCl_4$, and SiH_2Cl_2 (basic reagent). Because of the thermal decomposition of SiH_4 at temperatures above 520°C [40], the research concerning the feasibility of realization of the ALD mode was done only for $SiCl_4$ and SiH_2Cl_2 [41–45]. Some process data are presented in Table 1.

The authors of the work [41] did their research at 427 – 627°C in a high-vacuum individual chamber at different temperatures T_d and pressures P_d. The substrate of Si(100) was cleaned by methanol, acetone, and de-ionized water, then it was blown by the air cleaned from the microparticles, warmed up in a reactor at 602°C for 1 minute, and then it was subjected to the influence of H_2O plasma for removal of carbon, which as a result made it possible to receive SiO_2 with a thickness of about 2 nm and with a hydroxylated surface (SiOH*). Then, at the desired temperature of the substrate, a dose of $SiCl_4$ was injected $\geq 2\times10^9$ L (L – Langmuir unit of measure L = 10^{-6} Torr×s, the product of pressure P_d on the duration of the pulse of reagent t), the reaction of which with the surface is described by the following scheme:

$$SiOH^* + SiCl_4 \rightarrow SiOSiCl_3^* + HCl, \quad (5)$$

here and below the surface groups are designated with asterisks. During the next injection of ammonia, the dose ≥1.8×1010 L, the following reaction was assumed:

$$SiOSiCl_3^* + NH_3 \rightarrow SiOSiH_2^* + HCl \quad (6)$$

After several consecutive cycles of injection of $SiCl_4$ and NH_3, the surface was transformed from SiO_2 into Si_3N_4 (the corresponding schemes are not presented). This, according to the authors, made it possible to carry out the growth research on the in-situ generated S-SNTF. As an explanation of the schemes of the process, the authors assumed the course of two reactions (7,8) at the corresponding pulses of $SiCl_4$ and NH_3:

$$SiNH^* + SiCl_4 \rightarrow SiNSiSl_3^* + HCl \quad (7)$$

$$SiCl^* + NH_3 \rightarrow SiNH_2^* + HCl \quad (8)$$

At the parameters presented in Table 1, S-SNTF were formed without a nucleation delay by the linear dependence on the number of the cycles of deposition with the constant rate W_d, with the refraction index RI = 2.01 ± 0.05 and with a very small surface roughness (RMS) equal to 0.3 nm. The authors explained the twice as little Wd at the lower temperatures by a lower reactionary ability of the surface. Probably, this corresponds to the transition from region 1 to region 3 in Figure 5(a). Research [42] was done in an experimental reactor of a small size. The data concerning the preparation of the substrate surface are not available. Specific features of the research are the use of different T_d during injection of $SiCl_4$ and NH_3 and intermediate high-temperature annealing (900 – 1050°C) between the moments of injection of the reagents. The annealing revealed selectivity of the deposition rate on the surfaces of Si and SiO_2: the correlation was 2:1 at 900°C and 20:1 at 1050°C. The effect of annealing was explained by SiO_2^* and the adsorbed atom of Si* according to the scheme (9):

$$SiO_2^* + Si^* \rightarrow 2\ SiO \text{ (volatile product)} \quad (9)$$

The films were non-stoichiometric, the concentration of oxygen (hereinafter, [O], at %) was about 20%, the residual gases or impurities in the initial reagents were considered as the source of oxygen.

The topic was developed in several publications [43–48], the authors of which studied the transistor structures with SNTF on Si(100) substrates doped with boron with the resistance of 15 Ω×cm [44, 46] and 10 Ω×cm [43, 45]. Proceeding from a brief review of the totality of the data from these publications on the deposition process we will note the following:

- data on the used equipment are not presented;
- processing of the surfaces was done in the solution of NH_4OH:H_2O_2:H_2O with the proportion of 0.15:3:7 at 80°C during 10 minutes and the subsequent saturation (so-called termination) of the surface with hydrogen in the solution of HF (hereinafter DHF) diluted by water down to 0.5% for prevention of formation of a native oxide;
- injection of the reactants was done with the parameters specified in Table 1;
- the thickness of the films was about 0.4 nm [43,47], 3.5 nm (20 cycles) [44] and 4.5 nm (30 cycles) [48];
- after the process, the films were annealed in NH_3 at 550 – 600°C for 1.5 h.

Growth of TF on the substrates of Ge(100) [49], preliminary processed in DHF and annealed in NH_3 at 500°C during 10 min just before the deposition process, demonstrated a very low deposition rate (Table 1).

Comparison of the efficiency of the reagents of $SiCl_4$ and SiH_2Cl_2 with the use of NH_3 was done in the work [50] in the volume reactors, Figure 2(b), on the Si(100) substrates doped with boron and preliminary processed by BOE (Buffered Oxide Etchant). The deposition rate of the non-stoichiometric TF depended linearly, without saturation, on the doses of the reagents within intervals of $(0.2 – 5.8)\times10^{10}$ L and $(0.8 – 3.2)\times10^{9}$ L for pairs of $SiCl_4$ – NH_3 and SiH_2Cl_2 – NH_3, accordingly, reaching the values specified in Table 1. The higher rates of dissolution of the thin films obtained from SiH_2Cl_2 – NH_3 were connected with the higher concentration of hydrogen ([H], at %) in them because of its presence in SiH_2Cl_2.

Table 1. Brief results of SNTF research in cyclic TA-ALD reactors

Ref. ##	Reactants	T_d (°C)	P_d (Pa)	t_e (s)	W_d (nm/cycle)	Si:N
[41]	$SiCl_4$, NH_3	427–627	13–133	-	0.245	1:1.35
[42]	$SiCl_4$	375	25500	30–180	0.19 (на Si)	1:0.8
	NH_3	550	66500	120		
[43-48]	$SiCl_4$	375	22600	305	-	-
	NH_3	550	39900	90		
[49]	$SiCl_4$	350–400	23000	-	0.055	-
	NH_3		44000	-		
[50]	$SiCl_4$, NH_3	500	-	-	0.13	1:1
	SiH_2Cl_2, NH_3				0.12	
[51]	SiH_2Cl_2, NH_3	450	-	-	0.05	-
[52]	SiH_2Cl_2	200–400	8	90	0.09	1:1
	NH_3		24	15		
[53]	SiH_2Cl_2	375	8	90	0.1	1:1
	NH_3		24	20		
[54]	Si_2Cl_6	525–650	133	15	0.23	1:1.39
	N_2H_4		13.3	720		
[55]	Si_2Cl_6, NH_3	443–493	133–1330	-	0.23	1:1
[56]	Si_3Cl_8, N_2, NH_3	410–450	-	-	-	
[57]	Si_3Cl_8	310–500	93	2–15	0.027	1:1.3
	NH_3			1–30		

In the work [51] the same authors studied the system of SiH_2Cl_2 – NH_3 and considered the structure and properties of TF within the range of 450 – 550°C in more detail. The deposition rate resulted in the saturation with the value of about 0.08 nm/cycle took place only at 450°C and the doses of the reagents of $(0.6 – 6)\times10^{10}$ L. However, such TF turned out to be 40 times more quick-dissolving in DHF as compared to S-SNTF and unstable in time. During the two days of exposition to the open air, a 20% growth of the film thickness and lowering of the value of RI from 2.3 down to 1.8 were observed. The films grown at higher temperatures also turned out to be more stable and resistant to solutions of DHF. However, at $T_d > 450$°C and all the studied doses of the reagents (from 0.3×10^{10} L up to 6×10^{10} L) there was no saturation testifying in favor of the ALD mode, while the value of W reached 0.4 – 0.45 nm/cycle, exceeding the monolayer considerably. In conclusion, we will mention the first by time and the most frequently cited works [52, 53], although they present not a strictly thermally activated reception of TF from dichlorosilane. In the work [52] the system of SiH_2Cl_2 – NH_3 was used at a temperature of 200 – 550°C, however, the ammonia, injected into the reactor by pulses as the first reagent, was subjected to a preliminary remotely plasma-activated decomposition at the frequency of 2.45 GHz and plasma power of 30 – 60 W. Before the deposition the substrates of Si(100) doped with boron and resistance of 10 Ω×cm were first processed in the solution of H_2O:H_2O_2:NH_4OH = 100:20:1 with the subsequent immersion in DHF and rinsing in the de-ionized water within 1 minute. Such processing made it possible to terminate completely the surface with hydrogen. Certain parameters of the processes are presented in Table 1. The duration of the exposition in dichlorosilane was 45 – 450 s, the optimal was 90 s. The reaction resulted in saturation with the rate of about 0.09 nm/cycle in most of the experimentally verified conditions within the range of temperatures of 200 – 400°C. At high temperatures, the rate of deposition started to increase sharply. The non-stoichiometric TF had RI = 1.6 – 1.7; [O] ≈ 7 at % (which was explained by a possible dispersion of the quartz walls of the generator of plasma), [H] ≈ 13 at %, chlorine ([Cl]) ≈ 0.5 at %.

The authors of the work [53] investigated the deposition of SNTF at 375°C with the use of a forced activation on a heated up to 1700°C tungsten tape of the pulse injected ammonia (the method is called Catalytic CVD, Cat-CVD, or Hot-Wire CVD, HWCVD). At that, the pulse of the partially dissociated ammonia was the first. The substrates doped with boron Si(100) were processed in DHF for termination of the surface by hydrogen; for comparison, the substrates and S-SNTF grown at CVD were used. The basic results are presented above, in Table 1. Growth rate saturation was observed during the pulses of dichlorosilane over 90 s and of the dissociated ammonia – over 20 s. The authors discovered a considerably different growth of the films on the silicon substrates (linear growth of the number of cycles) and silicon nitride, on which a prolonged (~ 30 cycles) incubation period (nucleation delay) took place, explained by a lack of Si–H_x groups on the S-SNTF surface, which are available on the terminated silicon surface. The concentrations of oxygen and chlorine in the film were about 2 at %. The refraction index was about 1.85, the wet etch rate in HF solutions approached that of S-SNTF.

3.3. Cyclic Deposition with the Use of New Reagents

The new substances for deposition of the nitride of silicon became the oligomers of Si_2Cl_6 [54, 55], Si_3H_8 [56], and Si_3Cl_8 [57].

The authors of [54] studied the growth of TF in an experimental reactor with the participation of Si_2Cl_6 and hydrazine N_2H_4 at the temperature of 427 – 700°C (optimal – 600°C). The hydrazine was supplied first for the termination of the surface. According to the authors, the realization of the ALD mode (saturation by both reagents) took place at $P_d < 133$ Pa. At $P_d >$ 133 Pa the rate of the deposition increased twice and over, which was explained by the reaction in the gas phase of the remains of the reagents. In SNTF, close by composition to the stoichiometric ones, [O] = 4 at % was found (explained by the presence in the hydrazine with the cleanliness of 98% of the traces of water), as well as [H] < 1 at %, [Cl] – below the detection limit.

The authors of [55] studied the Si_2Cl_6 – NH_3 system in a reactor with cold walls at the temperature of the process of 515 – 573°C and doses of the reagents of $(0.1 - 3)\times10^8$ L for Si_2Cl_6 and $(1 - 10)\times10^8$ L for NH_3 (see the Table 1). It was revealed, that the real temperature of substrate Td measured by the calibrated thermocouples, in case of a pulse injection of the reagents is essentially less than the set "process temperature": 453, 496 and 513°C against 515, 557 and 573°C, accordingly. Si(100) substrates doped with boron and processed in BOE with the subsequent washing in the de-ionized water were used. Saturation of the deposition rate took place at doses > 1×10^8 L for both reagents. At $T_d > 513$°C the deposition rate increased sharply, corresponding to the transition from region 3 to region 5, see Figure 5(a). According to the authors, at $T_d > 500$°C, the deposition of TF goes in two directions: ALD and nitridation of the silicon TF – a product of the thermal decomposition of Si_2Cl_6 (its beginning is estimated at 500°C). The nonstoichiometric films had RI = 1.7 – 1.8, [O] = ~ 8 at % (the result of oxidation by the air after the process), [Cl] – below the detection limit.

In the work [56] with reference to the industrial volume low pressure reactor with hot walls, Figure 2(b), an example of a synthesis of the films was considered at a low temperature and cyclic organization of the process with the use of Si_3H_8. The process of Wd < 0.01 nm/cycle was not an ALD as such. It was carried out by the thermal decomposition of Si_3H_8 with the subsequent conversion of the formed amorphous silicon into the silicon nitride using several versions of activation by plasma of the nitrogen-containing reagents. The refraction indexes for the films enriched with silicon were: for N_2 RI = 2.1; for N_2+NH_3 (low concentration) RI = 2.18; for N_2 + NH_3 (high concentration) RI = 2.77.

The authors of [57] investigated the Si_3Cl_8 – NH_3 system (see conditions in Table 1) on the substrates with the structures with the aspect ratio equal to 60 and native silicon oxide without processing. In the range of the investigated doses of the reagents, no saturation was discovered. An inverse dependence of W_d on T_d was observed: from 0.6 nm/cycle at 310°C with a resulting constant value of about 0.027 nm/cycle at 440 – 500°C. That kind of dependence, probably, corresponds to the transition from region 2 to region 3 in Figure 5(a). This is confirmed, in particular, by complete oxidation in the air of TF obtained at 310°C and a sharp increase of the rate of dissolution in DHF. The main conclusion from the work is that the growth of TF proceeds according to the CVD reaction. At the temperatures of obtaining below 400°C, TF were very non-stoichiometric with [O] up to 65 at %; at T_d = 500°C [O] = 20 at %, [Cl] = 2 at %. The conformality of the deposition was estimated at the level of > 80%.

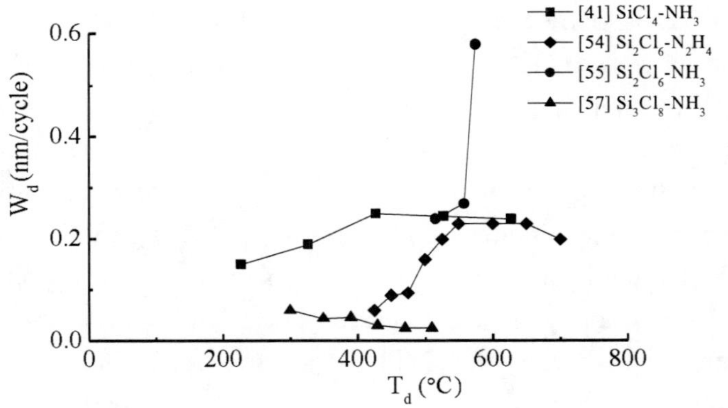

Figure 6. W_d vs T_d dependences for thermally ativated processes in cyclic reactors with the use of silicon chlorides.

Figure 6 shows W_d vs T_d dependences for thermally activated processes with the use of silicon chlorides. One can see relatively that dependencies are placed above ~ 450°C. In the case of the Si_2Cl_6 – NH_3 system, there are no ALD features that can be seen.

3.4. Conclusion to Section 3

The main direction in the development of the technologies for obtaining SNTF at a thermal activation of the reagents in the cycle type reactors is a decrease of the temperature of the processes by 200 – 300°C compared with the industrial processes for obtaining of S-SNTF in the flow-type reactors. With a view of studying the feasibility of realization of the atomic-layer deposition modes, the processes were investigated with the use of the industrial reagents of $SiCl_4$, SiH_2Cl_2, NH_3, and also new initial silicon-containing reagents-oligomers of Si_2Cl_6, Si_3Cl_8. Unfortunately, rare publications present full kinetic data, based on which it would be possible to judge the realization of the ALD modes. From the analyzed publications, a growth of SNTF in ALD mode can be considered possible in the works [41, 52, 54]. An essentially important condition for the growth in such modes is special attention to the aspects of preparation of the substrate surface for the elimination of the incubation period during the deposition. A specific feature of the processes is the necessity for large pulse doses of the reagents (high pressure or long duration of a pulse), which is not acceptable for the industrial use of such processes. At that, the authors of certain works pointed out the influence of a pulse on the substrate temperature during the deposition in the reactor, which should be taken into account in the development of the techniques of TF deposition. Judging by the presented data, in the other publications, the growth of SNTF was present, even during the cyclic realization of the processes, most likely, according to the mechanisms of the CVD reactions with participation of the substrate surface. In most of the research works the composition of the obtained films was non-stoichiometric in favor of the enrichment with silicon. The composition of the low-temperature films included oxygen in essential concentrations, which resulted in the refraction index at the level of 1.6 – 1.8. By the totality of the results presented in Table 1, for obtaining SNTF close to stoichiometry the deposition temperature should be not less than 500°C. However, at that, it is necessary to consider the possibility of the course of the reactions of the thermal decomposition of the initial silicon-contained reagents with nitridation of the formed amorphous silicon. At the moment, there are no available data about a radical improvement of the conformality of the deposition by ALD methods for SNTF.

4. Plasma-Actiivated Processes in Flow-Type Reactors

4.1. Brief Description of Thin Film Plasma-Activated Processes

As shown above, thermally-activated CVD processes are realized at T_d > ~ 600°C. This temperature is not applicable for the IC BEOL sub-route. A solution to the problem of the low-temperature TF became the use of the Plasma-Enhanced Chemical Vapor Deposition (PECVD) [1, 58, 59]. PECVD is called also "Plasma-Assisted CVD." For the first time, the PECVD processes were realized in simple-design "diode type" flow reactors with cold walls. An example of an individual reactor is shown in Figure 7(a), which is also called the reactors with the capacitance-coupled plasma, CCP. Besides, a frequently used term is parallel plate reactors. Later, for semiconductor technology adopted the tubular hot-wall LPCVD reactor, the multielectrode systems of a diode type were developed, Figure 7(b). In the case of PECVD at pressures P_d from several tens up to several hundreds of pascals, the deposition rates up to several hundreds of nm/min are reached. However, the temperature dependence of the deposition rate is normally very weak. At that, the values of the density of radiofrequency (RF) power per unit of the surface of a semiconductor substrate (W/cm^2) have a strong impact on the deposition rate and properties of TF. The plasma methods for obtaining TF are not strictly chemical processes and there are no experimentally proved gross reactions for them.

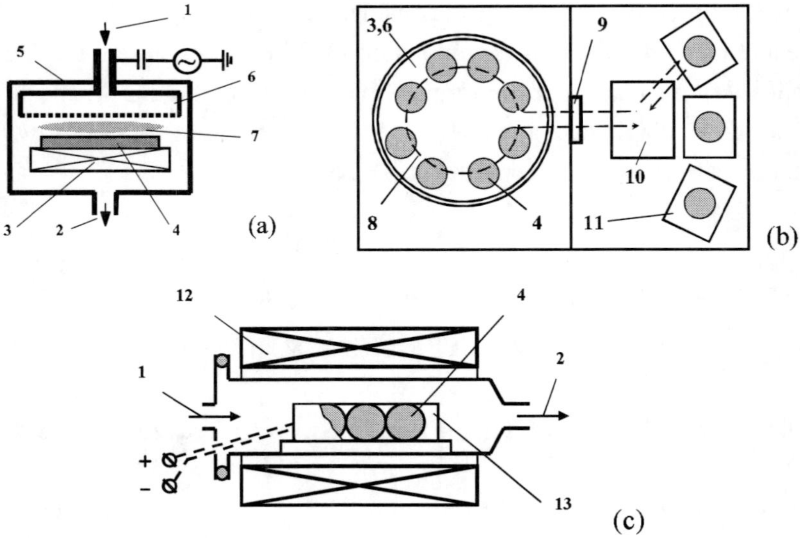

Figure 7. Simplified sketches of the most common PECVD reactors: cold-wall single-wafer PECVD chamber (a); cold-wall multy-stage PECVD (b); hot-wall multi-wafer tubular PECVD chamber (c). Definitions: 1 – reactants “in”"; 2 – reaction by-products “out”; 3 – heater/RF electrode; 4 – semiconductor substrate; 5 – reactor; 6 – gas mixture assembly or showerhead/ RF electrode; 7 – plasma area; 8 – wafer moving direction through six deposition stations with substrates; 9 – vacuum loadlock; 10 – robot system; 11 – cassettes with substrates; 12– furnace heater; 13 – electrode system with substrates.

In the case of a low-temperature PECVD with SiH_4, NH_3, N_2, the non-stoichiometric silicon-nitrogen-hydrogen-containing thin films (SiNH-TF) are formed. Their composition depends on the temperature, the nature of the reagents, composition of the gas mixture, RF power density, remoteness of a plasma source from a substrate, frequency of the RF discharge, etc. [1, 5]. Such films are designated also as $Si_xN_yH_z$, α-SiN_x:H, α-SiN:H [8-10]. The main undesirable component of SiNH-TF is hydrogen (usually, it is presented as a set of Si–H, N–H bonds), reaching the total concentration [H] up to 40 at %, thus it deteriorates such technologically important properties of SiNH-TF as density and rate of dissolution in the liquid etchants (wet etch rate, V_{etch}). The PECVD processes are usually used, for example, in admitting only low temperatures technological sub-route for manufacturing of IC, known as BEOL. Besides the non-stoichiometric composition, a principal drawback of all the PECVD processes of the thin films is the bad (20 – 60%) conformality of the films on a complex relief of the devices [2]. The bad conformality of TF complicates the applicability of SiNH-TF in the perspective technologies and devices with the relief structures. However, it is a motivator for the search for new solutions, because the interest exactly in the low-temperature technologies for obtaining the conformal SiNH-TF grows with the appearance of new technological problems.

In this section, an analysis is performed of the directions for the development of the technologies for obtaining low-temperature PECVD of SiNH-TF in the flow reactors (and also information concerning the composition, structure, and physical and chemical properties of these films). The literature on the given question is extensive and in most cases, it concerns the research of applicability of SiNH-TF in various technologies; including certain survey publications regarding ICs and MEMS [60–64], and also the works of direct comparison of the processes and coatings of S-SNTF and SiNH-TF [6, 7]. In this section, we consider only the experimental results about the "classical" reactors of PECVD with CCP. The use of the so-called "high-density plasma" (HDP) at PECVD will be analyzed in the next section of this Chapter.

4.2. Previous Summaries of Publications Concerning PECVD SiNH-TF

The first review materials on SiNH-TH were published in the mid-late 1980s [1, 60, 61]. The research was mainly done in the reactors with cold walls and CCP, for example, in the reactor of an individual type, Figure 7(a). For deposition of SiNH-TH at pressures of 30 – 300 Pa the plasma with a frequency of 50 – 450 Hz and 13.56 MHz was used; in terms of a unit of the substrate surface, the density of power was 0.02 – 0.2 W/cm^2. The primary goals of research and perfection of SiNH-TF as material regarding the IC technologies were: a

- studying of a possibility of decreasing of PECVD temperature and properties of coatings with preservation of the high deposition rates;
- search for ways to decrease and fix the level of the total concentration of hydrogen [H] in SiNH-TF;
- increase of the density of SiNH-TF and decrease of the rate of etching (that is, for a practical application – improvement of the selectivity of etching to the silicon dioxide films). The review [61] systematizes the data about the PECVD processes and properties of SNTF received with the use of inorganic reagents of SiH_4, SiF_4, $SiCl_4$, SiI_4 (the main of which is monosilane) in the presence of ammonia and nitrogen. It generalizes the literature data on the deposition rates W_d, the composition of the films in terms of Si/N ratio, the total concentration of hydrogen [H], concentrations of [Si–H] and [N–H] bonds, dependencies of the refraction index RI, density (γ), etch rate in liquid etchants V_{etch}, mechanical stresses (σ) during variation of the characteristics of the processes (temperature, pressure, fractions of the nitrogen-containing components in a gas mix, the density of RF power, frequency of RF generators, etc.).

The results of the detailed generalizations [61] are presented in Table 2, where the forms of the arrows graphically display the directions of the responses of characteristics of SiNH-TF in case of an increase of the parameters of the investigated PECVD processes specified in the table. Besides, this work presents a generalization of the data of the series of reports concerning the feasibility of obtaining SiNH-TF at PECVD from less dangerous, than monosilane, initial organic silicon compounds, for example, hexamethyldisilazane $[(CH_3)_3Si]_2NH$, hexamethylcyclotrisilazane $C_6H_{21}Si_3N_3$ and diethylaminotrimetilsilane $(CH_3)_3SiN(C_2H_5)_2$. Out of the publications analyzed in [61], it is necessary to mention two publications supplementing each other.

In [65] at a fixed temperature of deposition of 300°C, the pressure of 25 Pa and density of RF power of 0.15 W/cm^2 in the range of correlations of flows of NH_3/SiH_4 of 1.33 – 4.44 the variations of the Vetch in BOE solution, RI, the correlations of Si/N ratio and SiNH-TF resistance were studied. It was established, that the Si/N ratio varied within the limits of 0.95 – 1.25, total concentration [H] = 20 – 25 at %, at that, the author recognized that the most important influence on SiNH-TF characteristics was rendered by the concentration of Si–H bonds.

Table 2. Graphic responses of SiNH-TF characteristics on the PECVD process parameters increase for SiH_4–NH_3–N_2 reactants

PECVD	Thin film parameters response (based on [61])								
parameters	W_d	Si/N	[H]	[N–H]	[Si–H]	RI	γ	V_{etch}	σ
T_d	↗	↘	↘	↘	→	↗	↗	↘	↘
P_d	↗	↗	→	↘	↗	↗	↘	–	↘
Nitrogen compounds share in gas mixture	↘	↘	↗	↗	↘	↘	↗↘	–	–
RF power density	↗	↘	→	↗	↘	↘	↗	–	–
RF frequency	↘↗	→	↗	–	–	↗	↘	↗	↗

In [66] a comparison was done of the concentration of hydrogen and the rate of dissolution for 19 various versions of the industrial processes of PECVD SiNH-TF realized on various installations of the parallel-plate capacitance-coupled plasma type from various world manufacturers, and also experimental research was done of the SiH_4–NH_3–N_2 system. The author made the following conclusions:

- for mixes of SiH_4–NH_3 the maximal level of hydrogen in SiNH-TF was [H] = 39 at % at 150°C and 20 at % at 380°C, but it was approximately by 10 at % less than at the same temperatures for mixtures of SiH_4–N_2;

- hydrogen was present in SiNH-TF mainly in the bound kind as a set of Si–H and N–H bonds, and its general concentration in SiNH-TF exponentially correlated with the values of the etch rate BOE (the author of the review evaluates this correlation as follows: $V_{etch} \sim 0.99^{0.23[H]}$.

4.3. Industrial Technologies for the Production of SiNH-TF by PECVD

Table 3 presents the average data from [1] with additions from [6, 7, 61, 62, 66] on close to the industrial processes of S-SNTF and SiNH-TF. The data concerning the basic characteristics of the films are presented as of the end of the 1990s regarding the IC technologies. The survey work [6] presents a detailed table of important SNTF characteristics for application in MEMS with an indication of the primary sources of the information. In the other works, for example in [67], it is possible to find the comparative data on the important parameter of the films – hardness: for S-SNTF – ~35 GPa, and SiNH-TF, in a wide range of conditions of obtaining – from ~12 up to ~22 GPa; at that, according to the data [68], fast thermal annealing of SiNH-TF at 800°C allowed to approximately double their hardness. The RI and σ are the routine-controlled parameters in manufacture among the number of the SiNH-TF controlled parameters. For a fast analysis of the concentration of hydrogen in SiNH-TF in the industry they are deposited on the substrates of monolithic silicon (including the semiconductor substrates of large diameter) and analyzed by Fourier transform infrared, FTIR, non-destructive method). In FTIR spectra of SiNH-TF, besides the peak of Si–N of ~850 cm^{-1}, there are peaks of Si–H at ~2160 cm^{-1}, and of N–H at ~3350 cm^{-1} [5, 62, 69].

4.4. Directions for Improvement of the Technologies of SiNH-TF by PECVD

The analysis allowed us to structure the following directions for the development of the PECVD processes in the flow reactors.

Direction 1 characterizes optimization of the PECVD processes in industrial reactors concerning a concrete purpose of SiNH-TF in IC technology. For example, in [70] for application of SiNH-TF as an intermediate layer under the film of the interlayer dielectric of borophosphosilicate glass (BPSG) the versions of the parameters of the PECVD process were considered for optimization of the values of the mechanical stresses and elimination of cracking after high-temperature vitrification of BPSG. To the same direction, we can also refer the information [71] about the technology for obtaining PECVD SiNH-TF with the use of monosilane on the production PECVD tool, allowing to vary the values of the mechanical stresses in the deposited film from –3.5 GPa (compressive) up to +1.7 GPa (tensile) with an acceptable conformality of deposition. Unfortunately, detailed information concerning the composition of SiNH-TF is not available. It is possible to assume, that the values of the mechanical stresses vary employing the film enrichment with silicon, which was considered in [12] for the case of high-temperature films.

In [72] a version is considered for realization of PECVD SiNH-TF from SiH_4–NH_3–N_2 mixture at the temperature raised by ~200°C in comparison with the traditional PECVD. In the FEOL sub-route of ICs, such films replaced S-SNTF, received at temperatures above 700°C in hot-wall LPCVD. Raised to 580°C temperature of PECVD with the process of comparison of PECVD at 380°C for the RF power of 170 W with the frequency of 13.56 MHz allowed to lower the concentration of bonds [Si–H] from ~4.5×10^{21} down to ~4.0×10^{21}) at/cm^3, and for the bonds [N–H] from 6.5×10^{21} down to 3.0×10^{21} at/cm^3 (the measurements were made by FTIR method). The etch rate appeared to be twice as low, while the values of the compressing mechanical stresses were recorded at the level of 1.55×10^9 dynes/cm^2 (equal to –155 MPa). In [73] similar SiNH-TF were characterized using the SIMS (secondary ion mass spectrometry). For SiNH-TF obtained at 550 and 380°C the total concentration of hydrogen was 1.6×10^{22} at/cm^3 (~16.9 at %) and 4.0×10^{22} at/cm^3 (~35 at %), accordingly, which corresponded to the data of the work [72].

Direction 2 includes the publications dedicated to a deeper research of the PECVD processes and SiNH-TF properties. For example, in [74] SiNH-TF were compared, received from SiH_4–N_2 mixtures with argon and hydrogen, in [75] an opportunity was considered to obtain SiNH-TF at 200 – 350°C with a lower content of hydrogen from SiH_4–NF_3–N_2 mixtures, in the works of [22, 76–79] the physical-chemical and mechanical properties of SiNH-TF were studied with the use of modern analytical technologies, in [73, 22, 80, 81] the electric characteristics of SiNH-TF were studied. In [82] utilizing FTIR method the author investigated the hydrogen loss from SiNH-TF, received in a remote plasma in excess of ammonia at 200°C at thermal annealing, which began from 500°C and was characterized as a result of the interaction of Si–H and N–H groups with formation of Si–N bonds and H_2. The author of [83] as the initial siliceous substance silicon-containing reagent for PECVD investigated disilane Si_2H_6, in [84] the films received by magnetron sputtering of silicon in the argon-nitrogen-hydrogen atmosphere were investigated, the authors of [85] considered the results of PECVD in CCP reactor with three electrodes.

Direction 3 includes the publications about the research of the conformality of PECVD SiNH-TF. The requirements to the conformality of coating of the relief by thin films increase in the process of development of the microelectronic technologies and complexity of the semiconductor structures. At the level of IC technologies of approximately 0.35 – 0.18 μm and less, the non-conformality of deposition leads to problems of formation of voidings and, accordingly, to a search for technological possibilities for a "gap-filling" in the IC relief without voidings [2, 86]. Since the conformality of deposition of coatings is often declared regarding relief of an arbitrary form, it is difficult enough to interpret quantitatively the results presented in

Table 3. Summarized data on the processes of LPCVD S-SNTF and PECVD SiNH-TF

Thin film definition	S-SNTF	SiNH-TF
Deposition method	Thermal LPCVD	PECVD
Reactor type	Hot-wall tubular	Cold-wall CCP
Reactants	SiH_2Cl_2, NH_3	SiH_4, NH_3, N_2
T_d (°C)	700–900	250–350
P_d (Pa)	40–65	33–270
W_d (nm/min)	2–6	30–150
Si/N	0.75	0.8–1.2
[H] (at.%)	4–8	< 40
RI	2.0±0,02	1.8–2.5
γ (g/cm^3)	2.9–3.1	2.4–2.8
σ ($\times 10^9$ dyne/cm^2)*	12–18(T)	1(C) –8(T)
Dielectric constant	6–7	6–9
H_2O permeability	Zero	Low/Non
Na^+ penetration depth, nm	<10	<10
Breakdown strength ($\times 10^6$ V/cm)	5–10	4–6
Specific resistance ($\Omega \times$cm)	10^{15}–10^{17}	10^6–10^{15}
Surface resistivity (Ω/square)	>10^{13}	10^{13}
Thermal expansion coefficient ($\times 10^{-6}$ K^{-1})	4	4–7
Thermal stability	Excellent	<400°C
V_{etch} (nm/min): - 6:1 BOE (20°C–25°C) - 49% HF (23°C)	 1.0–1.5 8,0	 20–35 150–300
V_{etch} in 85% H_3PO_4 (nm/min): - at 155°C - at 180°C	 1.5 12	 10–20 60–100
V_{etch} in 30% KOH at 70°C (nm/min)	0	0.68
Plasma etch rate in 70% CF_4/30% O_2, gas mixture at 150 W, 100°C (nm/min)	20	50
Thin film conformality (%)	~100	~20–60

Notes. *Mechanical stress definition: (T) - tensile, (C) – compressive. To convert dyne/cm^2 to Pa need to use the correlation: 1×10^7 dyne/cm^2 = 1 MPa.

the research, especially for the PECVD processes. However, for PECVD typical are the sections of the structures with a superfluous quantity of the deposited material on the top parts of the relief steps, Figure 8(a), which are often called "bread loaf." The conformality of obtaining coatings for the PECVD processes essentially concedes to practically an ideal conformality of S-SNTF, received due to the use of the thermally activated processes of deposition, Figure 8(b). In his works [2, 40] the author formulated rather simple methodical approaches and gave quantitative characteristics to various processes of deposition of the silicon-based coatings concerning the IC relief with the steps rectangular in section. One of the conclusions of the presented publications is the established quantitative interrelation of the conformality of deposition and an effective constant of the deposition rate of the CVD processes of thin silicon-based films (k_{eff}, cm/s). The latter is the relation of the deposition rate (W_d) to the concentration of a silicon-containing reagent in the gas phase [C_{Si}]: $k_{eff} \sim W_d/[C_{Si}]$. The experimentally found values of k_{eff} varied within three orders of magnitude [2]. It appeared possible to connect the conformality of the correlation for the investigated processes with the values of k_{eff} in the following way: $d_2/d_1(\%) \approx 45.4 \times k_{eff}^{-0.51}$. A practical consequence from the given expression is the conclusion about the impossibility of obtaining the conformal coatings in the activated processes for obtaining the films having great values of k_{eff}. For the PECVD processes of the dielectric films, the conformality is estimated as 20 – 60%. On the contrary, the slow processes of deposition with small values of k_{eff}, for example, the high-temperature process of S-SNTF in LPCVD (the first column of Table 3), gives practically 100% of conformality of deposition [72].

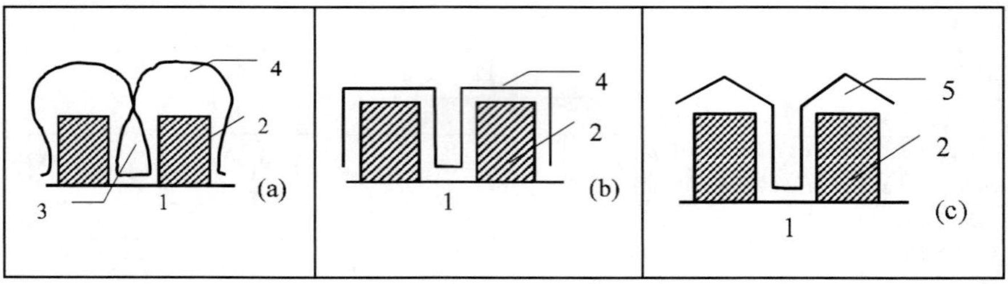

Figure 8. Simplified structure cross-section sketches of strictly non-conformal PECVD film (a), almost conformal LPCVD film (b), and specific biased HDP-CVD film: 1 – semiconductor substrate; 2 – conuctor lines; 3 – emptiness in between steps due to non-conformal deposition; 4 – conformal thin film; 5 – specific biased HDP-CVD film profile.

The author would like to underline, that, if in the literature the values of conformality for the plasma processes are higher than the ones specified above, this, as a rule, is explained by the application of the structures with the inclined lateral walls. There are few publications with the quantitative results of the research of the conformality of growth of SiNH-TF during a plasma activation. For example, in [72] the conformality for the processes of PECVD SiNH-TF at 380, 430, 480, 530, and 580°C was estimated approximately as 52, 54, 57, 64, and 69%, accordingly, however, the initial structures themselves are not characterized.

4.5. Conclusion to Section 4

The PECVD processes with the participation of the silicon-containing reagents, gaseous ammonia, and nitrogen in the flow reactors at temperatures below ~ 400°C are very important for IC device technology. Their advantages are a relative simplicity of realization, essentially smaller temperature dependences, and high deposition rates of films. The major problem of a low-temperature PECVD from SiH_4–NH_3 mixtures is the non-stoichiometry of SiNH-TF, in particular, an undesirable for the technology presence of hydrogen with the total concentration up to 40 at %, bound with the silicon and nitrogen, due to which the films have a low density and high rate of dissolution. In the following part of the review, we will present the results of the research of deposition of SiNH-TF in the reactors with high-density plasma, allowing us to overcome certain technological problems of PECVD.

5. Processes in Flow-Type Reactors with High Density Plasma

5.1. Brief Description of Thin Film High Density Plasma Activated Processes

Section 4 presents the state and directions for the development of the technologies for obtaining low-temperature non-stoichiometric SiNH-TF for

IC and MEMS with the use of PECVD. Chronologically the first PECVD processes of the thin films (TF) were realized in simple design CCP reactors (known as "parallel plate reactor," "direct plasma reactor"). The advantages of such reactors are the high growth rate of films at a low (<400°C) temperature of deposition, and versatility of applications for the substrates of different types and materials. Among the basic drawbacks of SiNH-TF, received using PECVD, are a non-stoichiometric composition (variation from the stoichiometric correlation of Si/N = 0.75 up to the values ~1.2) owing to enrichment with silicon, and presence of hydrogen with the total concentration up to 40 at %, connected with silicon and nitrogen in the form of Si–H and N–H bonds. Because of their non-stoichiometry, SiNH-TFs have a low density and high wet etch rates, which can complicate the use of the given materials in certain technological applications. Besides, in the case of PECVD, the conformality of deposition of SiNH-TF on complex reliefs is unsatisfactory, which complicates their applicability in the perspective technologies and devices with the relief structures. The typical properties of PECVD SiNH-TF are presented in section 4 in comparison with the stoichiometric high-temperature S-SNTF.

A solution to the above problems of SiNH-TF was found during the use of the PECVD method with activation of the so-called "High-Density Plasma CVD, HDP-CVD." The PECVD processes of dielectric TF with the use of HDP concerning the traditional methods of PECVD are characterized by a higher concentration of the active particles in plasma – by one or two orders, up to 10^{12} particles/cm^3. Studies of deposition HDP-CVD of TF in the experimental reactors were started back in the 1980s [2–7], and from the beginning of the 1990s, such reactors were applied actively in the industrial technology of ICs. The basic difference of deposition with HDP-CVD in comparison with the classical PECVD is a similarity of the characteristics of the deposited material to that of S-SNTF, both in respect of Si/N correlation and the content of hydrogen. To make them distinct from SiNH-TF in the present review such coatings are called HDP-SNTF.

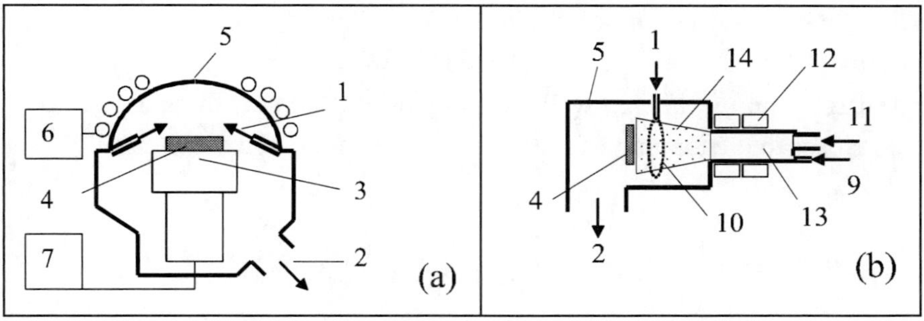

Figure 9. Simplified sketches of production flow-type single-wafer cold-wall reactors for deposition of HDP-SNTF: (a) inductively-coupled high-density plasma reactor; (b) electron-cyclotron high-density plasma reactor. Definitions: 1 – reactants "in"; 2 – reaction by-products "pump-out"; 3 – heater; 4 – semiconductor substrate; 6 – reactor; 7 – RF generator connected to an inductor; 8 – RF generator connected with an electrostatic chuck to produce bias for "in-situ" sputtering of deposited films; 9 – nitrogen-contained gas "in"; 10 – monosilane distribution ring; 11 – microwave power "in" (2.45 GHz); 12 – magnets; 13 – plasma cavity; 14 – plasma flux.

For the realization of the reactors with HDP-CVD, several types of excitation of the high-density plasma are used. Simplified sketches of the reactors with HDP-CVD are presented in Figure 9 (a,b). Accordingly, obtaining the coatings in the reactors with HDP-CVD of various types is designated below as ICP-HDP and ECR-HDP. Reactors with HDP are usually oriented on one substrate and can be used in a version with the so-called "remote plasma" when a generation of plasma is carried out beyond the placement of a sample. This allows reducing the undesirable influence of plasma on the substrate. A specific feature of PECVD with HDP is a low, only fractions of pascals, working pressure, and in this connection high-concentration reagents are applied (as a rule – monosilane SiH_4). The character of growth of TF at PECVD with HDP is similar to the traditional PECVD or even has a more non-conformal character than the traditional PECVD, see Figure 8(a). However, in the reactors with HDP, it is possible to achieve a radical improvement of the deposition conformality due to the use of a simultaneous with the deposition sputtering of the deposited material, for example, by argon ions, at an angle to the substrate [87]. This provides an opportunity to etch the top hanging surpluses of the material and to change the form of the growing TF, as is shown in Figure 8(c). At that, the growth of TF in a gap occurs due to the material coming from the gas phase, and also as a result of a repeated deposition (re-deposition) of the sprayed material on the gap walls. It is obvious, that such a re-deposition occurs closer to the top part of the gap, also worsening the conformality.

5.2. HDP-CVD SNTF: Basic Results

The references in [88–93] presented examples of the results of the research for the case of the use of the reactors c ECR-HDP and gas mixes of SiH_4–NH_3 and SiH_4–ND_3 [88], SiH_4–N_2–Ar [89,90], SiH_4–N_2 [91, 92)]. In [89, 92] a profound research results were presented, references [88, 92] were dedicated to studies of the concentration of hydrogen, and the work [93] – to the influence of the processes parameters on the values of the mechanical stresses in the films. Let us point out the following characteristics of the processes with ECR-HDP: frequency of plasma excitation – 2.45 GHz, power of the high frequency generator of plasma: 500 – 2500 W for the case of the deposition version remote from the generator [93] in the range of temperatures of 50 – 550°C and of 200 – 1000 W for the case of a direct impact of plasma [89–92)]. The level of the working pressure was 0.1 – 0.5 Pa, which was by several orders less than for a "classical" version of realization of PECVD. Depending on the level of high frequency power of the rate of deposition, in this case HDP-SNTF appeared to be at the level of 1 – 10 nm/min [89, 90]. The authors of the work [88] pointed out, that, in general, the total concentration of hydrogen was less, than that for PECVD, increasing from $\sim 6\times10^{21}$ (~5 at %) up to $\sim 2\times10^{22}$ at/cm^3 (~17 at %) at the decrease of temperature from 550°C down to 50°C. At that, the relation of the concentration of [Si–H]/[N–H] bonds for ECR version of deposition was ≥ 1.0 while for the comparative process of PECVD at 250°C it was <1. As an important result [88], we should underline, that by means of the method of the programmed thermal desorption the temperature of the beginning of diffusion of hydrogen from HDP-SNTF was established and was equal to 620°C. That was higher, than for SiNH-TF, received in the remote plasma at PECVD in excess of ammonia at the temperature of 200°C [82], the hydrogen diffusion in which began from 500°C and was characterized as a result of interaction of Si–H and N–H bonds with formation of the bond of Si–N and H_2.

The authors of the works [92, 93] compared directly the characteristics of the processes and SiNH-TF, received in a reactor with ECR at temperatures of 25, 200 and 300°C with those, received in the traditional LPCVD and PECVD; the results for 200°C are presented in the first column of Table 4.

Table 4. Summarized data on the process and parameters of S-SNTF and SiNH-TF

Thin film definition	HDP-SNTF		SiNH-TF	S-SNTF
Deposition method	ECR HDP	ICP HDP	PECVD	LPCVD
Reactants	SiH_4, N_2	SiH_4, N_2	SiH_4, NH_3, N_2	SiH_2Cl_2, NH_3
T_d (°C)	200	200	250–350	700–800
P_d (Pa)	~0.16	0.13–2.66	33–270	40–65
W_d (nm/min)	1.1	2–96	30–150	2–6
Si/N	0.71	0.66–1.0	0.8–1,2	0.75
[H] (at.%)	8.5	5.5–15	<40	4–8
RI	2.0	1.9–2,3	1.8–2.5	2.0±0,02
γ (g/cm^3)	2.87	-	2.4–2.8	2.9–3.1
σ ($\times 10^9$ dyne/cm^2)*	~4.5 (C)	2(C) –9(C)	1(C) –8(T)	12–18 (T)
Dielectric constant	7	-	6–9	6–7
Breakdown strength ($\times 10^6$ V/cm)	1.2	-	4–6	5–10
Specific resistance ($\Omega \times$cm)	5×10^{14}	-	10^6–10^{15}	10^{15}–10^{17}
V_{etch} (nm/min)	8	5	10–20	1.5

Notes. *Mechanical stress definition: (T) - tensile, (C) – compressive. To convert dyne/cm^2 to Pa need to use the correlation: 1×10^7 dyne/cm^2 = 1 MPa.

In [94–98] the work presents examples of publications concerning the research of the processes for obtaining HDP-SNTF in the reactors with ICP. The research was done in the range of temperatures of 25 – 400°C, pressures of 0.13 – 2.6 Pa, power of 200 – 2000 W for mixes of SiH_4–N_2–Ar, and also SiH_4–N_2–He [96]. For plasma excitation, the generators with frequencies of 300 – 400 KHz [95] and 13.56 MHz [97] were used. In the investigated ranges of conditions, the rates of deposition were 2.0 – 96 nm/min.

According to data [94], for a remote plasma source in the range of temperatures of 25 – 300°C the rate of deposition was 2.7 – 5.2 nm/min, Si/N ratio was 0.6 – 1.0, RI = 1.75 – 2.2, HDP-SNTF density varied within 2.4 – 2.9 g/cm^3. The total concentration of hydrogen was (3.0 – 5.5)×10^{21} at/cm^3 (2.6 – 8 at %), while during a decrease of the temperature of deposition from 300 down to 25°C it increased up to ~1.45×10^{22} at/cm^3 (~13 at %).

In the work [95], estimation of the properties of HDP-SNTF was done, as well as SIMS (secondary ion mass-spectrometry) profiling of S-SNTF and HDP-SNTF obtained in PECVD and ICP-HDP intending to find out the distribution of the atomic concentrations of Si, N, H in the thickness of the films. At 400°C, the pressure of 0.66 Pa, and the power of 2000 W, the rates of deposition were 96 nm/min, while the concentration of hydrogen did not exceed 5.5 at %. Estimations utilizing FTIR of the concentration of hydrogen in HDP-SNTF with RI = ~2.0 at the temperature of deposition of 200°C for the mix of N_2/SiH_4 were ~1.31×10^{22} at/cm^3 for N–H groups and ~7.67×10^{21} at/cm^3 for Si–H groups [98], which, as a whole, corresponding to the results of the other works. Average data on [95–97] are presented in the second column of Table 4. Comparison of the data presented in Table 4 on HDP-SNTF with the data for SiNH-TF and S-SNTF taken from [99] allows us to conclude that deposition with HDP can be realized at a considerably lower temperature (down to the room temperature) and it ensures a composition close to the stoichiometric one at a much smaller concentration of hydrogen. Among the drawbacks of the deposition with HDP, it is possible to mention an essentially smaller rate of deposition at very low working pressure. For the maintenance of such pressures, it is necessary to use expensive high-speed turbomolecular pumps. Because of its high cost, the HDP-CVD equipment should be applied only in absence of any other solutions, for example, for solving the problem of conformality of deposition and filling of IC relief of a complex configuration without cavities.

5.3. HDP-SNTF Deposition in Experimental Reactors

For obtaining SNTF with HDP several experimental versions of the reactors were used, the examples of which are presented in [100–102]:

- supersonic jet deposition was described in [100]. The essence of the process consists in obtaining a coating on a substrate surface as a result of the interaction of the separate jets of atomic silicon and nitrogen in a flow of helium, received from SiH_4 and N_2 using microwave frequency excitation. According to the authors, such a method of deposition allows receiving coatings from a controllable monolayer up to 1 nm/min. The authors received super thin (<4 nm) HDP-SNTF with the rates of 0.3 nm/min and investigated the electric characteristics of such coatings. According to Auger spectroscopy and FTIR, in the coating composition the silicon and nitrogen (Si/N ~0.8), and a small amount of hydrogen was discovered. Besides that, oxygen was discovered with an abnormally high concentration, nearby 15 at %, which, according to an assumption of the author, is a consequence of either insufficient vacuum, or is due to the oxygen coming from the quartz nozzles;
- high-temperature horizontal tubular flow LPCVD with a device for excitation of HDP at the input into a reactor called by authors of [101] as "HDP-assisted LPCVD." The design allowed the authors to implement deposition of SNTF with the use of dichlorosilane and ammonia on the substrates of 76 mm at a thermal activation at 720°C (LPCVD), at 300°C directly in a plasma device, and also at 720°C in the version using the remote plasma. It is obvious, that the additional activation led to a multiple increase of the rate of deposition in comparison with the growth rate at a thermal activation (1.5 nm/min). The authors had reasons to believe that a drawback of the process was a low RI ~1.7. Besides, the question remains unanswered about the uniformity of the growth of SNTF along the length of the zone of the tubular reactor, which, according to the author of the review, should be unsatisfactory;

- obtaining of HDP-SNTF at 400–500°C from $Si(CH_3)_4$ and NH_2CH_3 with the use of hydrogen radicals generated utilizing microwave radiation is proposed in [102]. The total concentration of hydrogen is estimated using FTIR at the level of ~2.2×10^{22} at/cm^3, and [Si–H] and [N–H] at the level of ~8×10^{21} at/cm^3, 1.4×10^{22} at/cm^3, accordingly.

5.4. Conformality and Directions of HDP-SNTF Deposition Improvement

Data presented in Table 4 demonstrate that HDP-SNTFs are an excellent solution for modern technological applications from the point of view of the quality of the deposited material at low temperatures. However, the relief structures of modern devices have a problem of conformal deposition of TF on the steps of the relief, which was considered in relation to the activated processes and material in [99]. For example, the typical PECVD "bread loaf" shape of TF on the structures can be seen, Figure 8(a). For HDP-CVD, for example, by the presented in [97(20)] micro photos of the chips of the cross-sections of the structures with HDP-SNTF for a rather simple edge structure with the height of the steps of the metal strips of 0.9 μm (at the visually inclined walls of the metal strips and the gap between them of 1.2 μm), an estimation of the conformality gives about 60%, which is not better, than for regular PECVD SiNH-TF. In general, the regularities of deposition with HDP-CVD qualitatively correspond to the established by the author interconnection between the conformality of the deposition and the effective constant of the rate of CVD processes of the thin silicon-containing films k_{eff} (dimension is cm/s), see discussion in [99].

However, during the use of HDP, new possible solutions were found for the problem of non-conformality of the deposited TF. Thus, the use of the additional in situ sputtering (etching of the deposited material), carried out at an angle to the vertical of the structures [87], allowed us to change considerably the character of formation and the final form of TF profiles on the relief, Figure 8(c). It turned out, that the form of such a profile can be controlled by changing the correlation of the rates of deposition and sputtering (deposition-to-sputtering ratio, D/S ratio). Unfortunately, in the literature there is not enough experimental data about filling of the gaps in the reliefs of the devices with HDP-SNTF, for example, some photos of the structures with HDP-SNTF, obtained at various D/S ratios, are presented in the work [97]. In the US patent application [103], according to the proposed solution, the deposition of HDP-SNTF on the relief substrates is supposed to be done not simultaneously with etching, but consecutively: first, the deposition, and then, after removal of all the reaction components from the working chamber, sputtering of the material from the top parts of the steps. According to the authors, the reason for such an approach is in the considerable compressing mechanical stresses of HDP-SNTF, capable to infringe the structures in the course of film creation.

5.5. Conclusion to Section 5

The directions for improvement of PECVD methods are connected with the use of the high-density plasma in the reactors with the inductive-coupled plasma, with the microwave frequency plasma based on the electron-cyclotron resonance, and also in the reactors with the microwave plasma. At the lower temperatures for obtaining with the use of SiH_4–N_2 gas mixes, the rates of growth are considerably less in comparison with the traditional PECVD. Deposition with HDP ensures many times smaller and acceptably low level of the content of hydrogen, essentially higher densities of films and lower rates of film dissolution, which makes such HDP-SNTF promising for the devices, not tolerating high temperatures during processing. The conformality of TF coatings for the plasma-activated processes concedes considerably to the conformality of S-SNTF, obtained due to the use of the thermally activated processes of deposition (about 100%). To improve the conformality of deposition and formation of the gaps filled with material in the relief of the device complex structures, the methods, and hardware of HDP-CVD are used, which combine the processes of deposition of thin films and in situ sputtering (etching) of the deposited material. The basic comparative data are presented about the processes and the parameters of the processes for obtaining films during the thermal activation, traditional PECVD, and HDP-CVD. The directions for improvement of the HDP-CVD processes for a better filling of the gaps in the relief of the devices were considered.

6. Plasma-Activated Processes in Cycle-Type Reactors

6.1. Brief Description of Plasma-Activated TF Deposition in Cycle-Type Processes

Section 3 contains an analysis of the technologies for obtaining SNTF in the cyclic-type reactors with a sequentially pulsed injection of the initial chemical reagents (called "precursors") during pulses of a certain duration, see Figure 4. In the mode of the atomic-layer deposition (ALD) at 400 – 650°C, the rate of growth W_d was less than one monolayer per one cycle of deposition, see Figure 6. This method allows obtaining SNTF close by their composition to stoichiometric Si_3N_4 (S-SNTF), which corresponds to the correlation Si/N = 0.75. In section 4 and section 5, an analysis was done of the present state and directions for the development of the technologies for obtaining low-temperature (<400°C) TF in the flow reactors with plasma activation. Plasma-enhanced chemical vapor deposition (PECVD) allows us to reach high TF growth rates. However, at a low temperature, the non-stoichiometric materials are formed named by the author as "silicon-nitrogen-hydrogen containing thin films" (SiNH-TF). The main technologically undesirable component of SiNH-TF is hydrogen with a total concentration of up to 40 at %, which worsens such technologically important properties of the films as density, wet etch rate, etc. Use of the HDP-CVD ensures many times lower and acceptable for the technology level of the content of hydrogen, an essentially higher density of the films and lower wet etch rates. However, an ineradicable problem with PECVD, including HDP, is an unacceptably low conformality of the low-temperature plasma-activated coatings (<50%) in comparison with that of the high-temperature S-SNTF, received at the thermally activated CVD (about 100%). For improvement of the conformality of deposition, biased HDP-CVD is used, combining the processes of deposition of TF and its simultaneous sputtering. Besides, a drawback of HDP-CVD is a strong undesirable influence of plasma on the devices.

For solving the above problems, the authors of several relatively recent publications, for example in [104–107], suggested using the low-temperature method of ALD with a "direct" or "remote" plasma activation of one of the reagents (hereinafter, PE-ALD). Reactors with PE-ALD are the reactors of the cyclic type, in which a pulsed injection of the reagents occurs according to the general ideology of ALD at that one of the reagents (in case of SNTF it is either ammonia or nitrogen) is injected into the reactor activated, Figure 10. The plasma discharge can be turned on either at once, in the cycle beginning, or after a while, after injection of a reagent; at that, the activated reagent can be injected first in the deposition cycle.

The subject of ALD is very popular in connection with a high demand for the method as a solution to the problems of increasing complexity involved in the development of TF technology. In the present section, an analysis is done of the publications on the subject of reception of activated ALD SNTF in the cycle-type reactors at a low temperature with the use of the initial substances of various nature. A feature of the publication activity for the case of ALD SNTF consists in the fact that the thermally activated (actually – high-temperature) ALD SNTF were mainly investigated in the first decade, while the plasma-activated ALD – in the second decade of the 2000s. In recent years several survey publications on the activated ALD SNTF [8, 9, 36, 38, 39] were published. Despite the availability of the industrial equipment for the PE-ALD, this method for obtaining SNTF, apparently, still cannot be considered a mature technology in comparison with obtaining S-SNTF in the flow LPCVD or SiNH-TF in PECVD reactors. The aim of this section is analysis and generalization of the experimental results concerning the opportunities and directions of the research, estimation of the possibilities of the method of the PE-ALD, the features of its hardware realization, and the basic results of the research in comparison with SNTF obtained by the other methods.

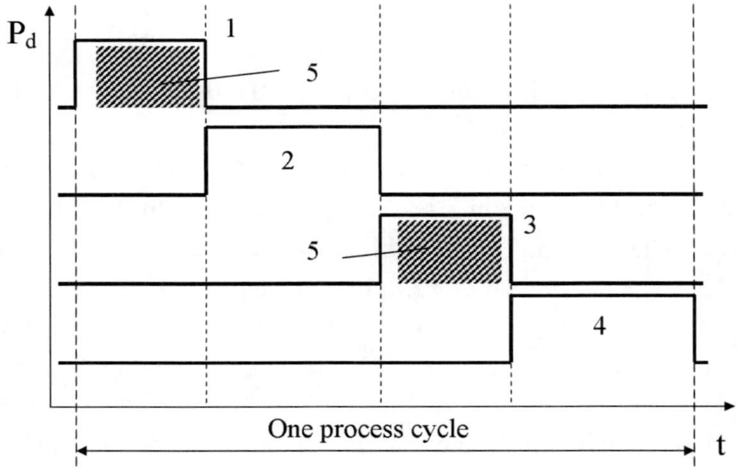

Figure 10. An example of typical plasma-activated ALD process diagram with reactant pulses and inert gas purge: 1 – precursor or nitrogen-containing reactant, 2 – purge gas, 3 – nitrogen-containing reactant or precursor; 5 – activation of nitrogen-containing reactant with plasma discharge either in pulse 1, or in pulse 3 with a little delay from the pulse beginning.

6.2. Brief Review of the Basic Types of the Reactors for PE-ALD

In general, the designs of the reactors for the PE-ALD are close to those of the plasma reactors, presented above and in Figure 11(a–d), and their detailed description and examples of the industrial reactors can be found in references [105, 106, 108]. As a rule, such reactors are the reactors of individual type (per one wafer). The first design of the reactor, Figure 11(a) is typical for a reactor with direct plasma, which envisages plasma chemical deposition from the vapor phase (PECVD) directly in the area of the substrate. In this case, the capacitance-coupled plasma (CCP) is usually generated with the frequency of 13.56 MHz between two parallel electrodes in the so-called "diode reactor." The gases are introduced into a reactor either through the showerhead or from one side, through the gap between the electrodes. Usually, the working pressure equals hundreds of pascals, but can also be about 10 times less. If an additional grid is placed in the area of plasma, Figure 11(b), we get a "triode design," allowing to keep plasma between the top electrode and the grid so, that the surface of deposition of the film will not contact the active plasma area. This somewhat reduces the density of the radicals and the ion flows to the substrate surface.

The third widespread configuration of a PE-ALD reactor presented in Figure 11(c) is called "ALD reactor with remote plasma." In this case, unlike in the reactors of the first type, for which concentration of the active particles can decrease down to zero in the direction to the wafer surface, here, plasma is present quite near the substrate surface, and thus the flow of the radicals to the wafer is much higher, than for the reactor of the previous type. In such systems, plasma can be formed like in reactors with the inductively coupled plasma (ICP), in reactors with a microwave frequency electron-cyclotron plasma (ECR), and microwave plasma reactors. It is important to point out, that the parameters of the plasma and the substrate in such reactors can change independently, expanding possibilities for variation of the characteristics of PE-ALD processes. The level of the working pressure Pd in such reactors equals some tens or hundreds of pascals. In the design of the PE-ALD reactor presented in Figure 11(d), the microwave source of plasma is connected to the reactor for CVD with a usual thermal activation of the wafer. This type of ALD plasma activation is called "radical-enhanced." Such systems are most frequently used as embedded in CVD/PECVD/PECVD installations with HDP for cleaning of the walls of the reactors after deposition by plasma etching.

The advantages of PE-ALD are [105] expansion of the experimental conditions for obtaining of the materials with the set properties, including due to expansion of the possibilities for selection of the precursors of the inorganic and organic nature, increase of the rate of deposition at the lowered temperatures (at temperatures less than 500°C they are considered the most promising), improvement of the properties of coatings, etc. Among the drawbacks are the problems with the conformality of coatings by films at PE-ALD on a developed relief of IC.

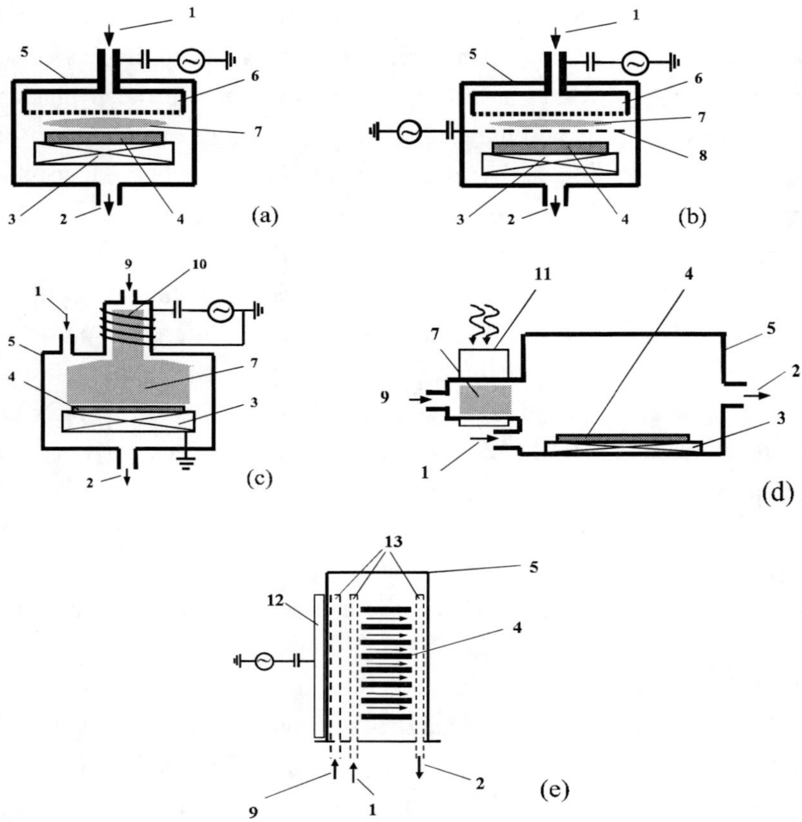

Figure 11. Simplified sketches of production flow reactors for stoichiometric silicon nitride deposition in hot-wall tubular CVD system with horizontal (a) and vertical (b) reactor orientation. Definitions: 1 – precursor (and reactants) "in"; 2 – reaction by-products "out"; 3 – heater; 4 – semiconductor substrate; 5 – reactor; 6 – gas mixture assembly or showerhead/ RF electrode; 7 – plasma area; 8 – grid; 9 – plasma gas "in"; 10 – coil on dielectric tube; 11 – microwave cavity; 12 – RF plasma for ammonia gas; 13 – perforated inserts for gas flow formation.

6.3. Thin Film Deposition with the Use of the Inorganic Precursors

The investigated precursors of the inorganic nature include monosilane SiH_4 [109, 110] and its derivative neopentasilane $(SiH_3)_4Si$ [111], trisilylamine $(SiH_3)_3N$ [111–115] and its derivatives [116], chlorine-containing silanes [52, 117–124]. Brief numerical data from the published research works are presented in Table 5.

6.3.1. PE-ALD from Silanes

PE-ALD from SiH4 was done on a silicon surface preliminary processed by plasma of nitrogen, on which, according to the authors, the surface terminated by nitrogen – a monolayer of nitrogen with active bonds, was formed. Saturation by the activated nitrogen took place at a pulse with a duration of more than 60 s. When SiH_4 is injected, it connects to nitrogen with the formation of Si–N bonds and N–H bonds. At the next injection cycle of the plasma-activated nitrogen, the N–H bonds are removed and replaced with Si–N bonds. The deposited films are enriched with silicon, at that, the mechanical stress of compressive type of –810 MPa, the density of 2.5 g/cm^3 and the RI = 1.75 were observed. For PE-ALD from neopentasilane [111] we should point out a rather long duration of the pulses of the plasma-activated nitrogen (Table 5), at that, even in case of the duration of 30 s the saturation of the rate of deposition is not reached (thus, the refraction index decreases from 2.14 down to 2.07). The films demonstrated a composition close to the stoichiometry, and small concentrations of carbon and oxygen (<1 and <3 at %, accordingly). The concentration of hydrogen was influenced by the power of plasma: for 275°C and plasma power of 250 and 750 W the concentration of hydrogen (at %) and the density of films (g/cm3) were 11 and 2.21; 23.5 and 1.86, accordingly. For the structures etched in silicon with the width of 0.25 – 2.0 μm and depth of 5.45 – 11.8 μm it was discovered, that the duration of a pulse of nitrogen and power of the plasma discharge did not influence the conformality of the deposition. The conformality can be considered unsatisfactory, because for example, for a rather simple structure with the width of 0.35 μm at the depth of only 1.3 μm (recalculated for an aspect relation, it is only 3.7) the conformality of the films was only about 31%.

6.3.2. PE-ALD from Amines

In [111–116] the authors from four various groups studied the possibility of obtaining SNTF with the use of amines and, first of all, trisilylamine

$(SiH_3)_3N$, which molecule includes an atom of nitrogen directly connected with atoms of silicon. According to [111–113)], the density of SNTF was about 2.8 g/cm^3, the wet etch rates were a little bit less than those for the regular SiNH-TF obtained by PECVD. Most interesting are the attempts to characterize the conformality of SNTF deposition on the transistor structures for the use of a film as a spacer. From the presented data it is visible, that despite the very transistor structures forms, their rapprochement per step from 240 down to 120 nm led to a deterioration of the conformality of PE-ALD SNTF by 20 – 30%. Nevertheless, that was essentially better in comparison with regular PECVD. The authors of the works [114, 115] established the optimal window of the process with an approximately constant rate within the range of 250 – 350°C (Table 5). By their composition, SNTF were close to the stoichiometric ones, although had the index of refraction of 1.75 – 1.8 and contained 13.4 – 8.1 at % of hydrogen, accordingly. It was established, that roughly a triple increase of power from 100 up to 300 W provided a chance to lower the wet etch rate of SNTF approximately three times. In [116] SNTF are investigated, obtained from the derivatives of trisilylamine (by replacement of SiH3 groups in the initial precursor by the groups of dimethylaminomethylsilyl or trimethylsilyl). Out of the three synthesized precursors, a reagent was chosen as a promising one for PE-ALD SNTF with the gross-formula ($C_9H_{29}N_3Si_3$): bis(dimethylaminomethylsilyl)trimethylsilylamine. For deposition a reactor of direct plasma was used, the temperature of deposition was 250 – 400°C, pressure – about 80 Pa, duration of pulses of the precursor – 1 – 9 s, and of the plasma-activated nitrogen – 4 – 15 s. In those modes, SNTF correlated Si/N of about a unit and RI = 1.97. The concentration of carbon in case of variation of the plasma power within 100 – 200 W and over increased within the limits of 0 – 5 at %. At that, the concentration of oxygen also increased within 7 – 11 at %, which the authors explained by oxidation of SNTF, enriched by silicon, in the ambient air after the process. For the structure with the width of 0.071 μm and the depth of 0.375 μm and with slightly inclined walls the conformality at the half of the depth was estimated at the level of 73%. In [8] it was also possible to find brief tabular data concerning the attempts to use for the PE-ALD the other organic aminosilanes in a direct or remote plasma of nitrogen and ammonia: bis(diethylamino)silane, tris(dimethylamino)silane, etc. The specific features of processes of PE-ALD from these precursors were unacceptably low rates of growth (100th fractions of a nanometer per cycle), and also the value of the RI=1.7 – 1.8.

Table 5. Results of Investigation of SNTF in cycle type of reactors with the use of inorganic precursors

Ref. ##	Reactants	Reactor type	T_d (°C)	P_d (Pa)	t_e (s)	W_d (nm/cycle)	Si/N
[109, 110]	SiH_4	CCP	250–400	130–520	0.5–5	0.025–0.2	1.2
	N_2				1–90		
[111]	$(SiH_3)_4Si$	CCP	250–300	25	1	0.14	0.88
	N_2			75	2–30		
[111]	$(SiH_3)_3N$	CCP	250–300	25	1	0.12	-
	N_2			75	2–30		
[112, 113]	$(SiH_3)_3N$	CCP	300–400	520	-	0.14–0.21	0.89
	N_2/H_2				-		
[114, 115]	$(SiH_3)_3N$	Remote	50–400	40	0.2	0.06	0.75
	NH_3	ICP			5		
[52, 117]	SiH_2Cl_2	Remote	200–400	8	90	0.09	1
	NH_3	microwave		24	20		
[118]	SiH_2Cl_2 NH_3, N_2, H_2	Remote ICP	350	-	-	0.024	0.8–1
[119]	SiH_2Cl_2 NH_3	RPBR	350–400	-	-	0.1	-
[120]	SiH_2Cl_2 NH_3	RPBR	350–500	-	-	-	-
[121, 122]	SiH_2Cl_2 NH_3	RPBR	400–600	-	-	-	-
[123]	Si_2Cl_6	CCP	350–450	~9	30	0.12	0.58
	NH_3				35		
[124]	Si_2Cl_6 NH_3, N_2	CCP	300	-	-	0.5–0.11	-

6.3.3. PE-ALD from Chlorosilanes

Chronologically, the industrial reagent of dichlorosilane (SiH_2Cl_2) was investigated first. Unfortunately, the data available in the literature concerning the growth processes at PE-ALD are very limited. In [52, 117] the authors present the results of the research of the films received in the experimental reactors of individual type with a remote source of plasma by nitridization of ammonia by plasma of a thin silicon film, grown at the thermal-activated pyrolysis of dichlorosilane at 750°C. At that, injection of the activated ammonia was the first in the cycle, followed by a pulse purge of the reactor by a mix of N_2–H_2 and then followed a pulse of dichlorosilane. The results of the research are presented in the table and in more detail in [38], where a direct comparison is done with the results of growth of SNTF at the thermal activation of the ALD process with the use of SiH_2Cl_2. Here we point out that the obtained SNTF had RI = 1.6 – 1.85, [O] ≈ 2 – 7 at % (which is explained by a possible sputtering of the quartz walls of the plasma generator), [H] ≈ 13 at %, [Cl] ~ 0.5 at %. The authors [118] compared SNTF, obtained at PA-ALD from SiH_2Cl_2, and the organic-silicon compounds with the use of the plasma-activated ammonia, nitrogen, and hydrogen. Certain data concerning the process with dichlorosilane are presented in Table 5, in more detail the results are considered below. A series of publications [119–122] present the results of the research done in the industrial batch reactors similar to LPCVD by their design, Figure 11(e), but working in the cyclic mode. Such reactors were additionally equipped with the devices of the remote high-frequency plasma activation of ammonia supplied to the reactor (Remote Plasma Batch Reactor, RPBR). The deposition was done in cycles from SiH_2Cl_2 at <500°C. The actual data concerning the process of growth of the films are very few, the numerical data about the processes are presented in Table 5. In the range of 350 – 500°C, the authors [120] found a linear variation of the density of films within the range of 2.5 – 2.75 g/cm^3 and a dome-shaped change of the compressive mechanical stress with a maximum of ~13 MPa at about 450°C.

In [121, 122] a possibility was estimated of the use of PE-ALD SNTF as a spacer material for IC technology with the technological nodes of 28 – 32 nm. The activated ammonia was also the first to be supplied in the deposition cycle. It was demonstrated, that the wet etch rates in DHF exponentially decreased with the growth of the temperature of deposition within the range of 400 – 600°C. The conformality of deposition on a transistor structure was also studied and it was equal to 100% for the unary and double pitches, while at the use of PECVD SiNH-TF the conformality was 60 and 87%, accordingly. Let us note an interesting result of the application of the precursor of monochlorosilane SiH_3Cl instead of dichlorosilane [122]. It turned out, that SNTF from this reagent had approximately twice smaller wet etch rates in DHF, which testified in favor of their higher density. Research works [123, 124] were done with the use of oligomer Si_2Cl_6 and plasma-activated ammonia. In the work [123] the dose of Si_2Cl_6 supplied to the reactor first was 8×10^5 L. For the temperature of deposition of 400°C in SiNH-TF the author discovered 23 at % of hydrogen, mainly in the form of N–H groups, RI = 1.9, density 2.5 g/cm^3. The presence of chlorine was estimated at a level less than 0.1 at %. The authors explained the film enrichment with nitrogen by the presence of N–H groups in it. The conformality of SNTF growth on the structure with the inclined walls and a gap in the bottom part of about 0.1 μm and height of 0.5 μm was practically ideal. The authors [124] used a flow reactor with cold walls (traveling-wave reactor, used in the very first research works of ALD [125]), additionally equipped with a plasma source. Besides Si_2Cl_6 precursor, the dose of which varied within the limits of 1×10^6 – 5×10^7 L, ammonia was used with the fixed dose of 1×10^9 L, and also a nitrogen plasma before injection of the precursor. Thus, the deposition cycle consisted of not the usual two stages, but of three stages: precursor/ammonia/activated nitrogen, at that, after each of the stages there was a purge by argon. Application of the nitrogen plasma (100 W, 200 Pa, 50 s) provided a chance to approximately double the rate of deposition from 0.05 up to 0.11 nm/cycle and, thus, to lower the dose of the precursor by almost an order.

6.4. Thin Film Deposition with the Use of the Organic Precursors

The whole set [118, 126–130] presents the results of the largest research works of the possibility of obtaining SNTF with the use of the PE-ALD from bis(tertiarybutylamino)silane, $(NH^tC_4H_9)_2SiH_2$, usually known as BTBAS.

A precursor is a colorless liquid with a density of 0.82 g/cm^3 at 20°C with the temperature of boiling of 167°C in normal conditions. For PE-ALD they usually use remote plasma of ammonia, nitrogen, and hydrogen in the reactor with ICP with the power of 600 W and frequency of excitation of 13.56 MHz. The following results of publications [126–130] may be regarded as the most interesting.

1. In [118], a comparative research was done of the correlation of SNTF density and wet etch rates in the hydrofluoric acid diluted with water in proportions 1:100. The films were obtained in a reactor with ICP, similar to the one described above, at the power of 300 – 400 W and deposition temperature of 350°C. As the initial organic precursor, tris(trimethyilamino)silane $[(CH_3)_2N]_3SiH$ was used, for comparison reasons, the inorganic precursors SiH_2Cl_2 and trisilylamine $(SiH_3)_3N$ were investigated. As the second nitrogen-containing reagent, the plasma-activated mixes of nitrogen, ammonia and hydrogen were used. In general, all the investigated versions of the precursors were described by a linear dependence within the range of wet etch rate of 0.2 – 9.0 nm/min and SNTF density of 2.9 – 2.3 g/cm^3. In all the types of SNTF no carbon was discovered, however, there was an abnormally high concentration of oxygen in it: from 7 up to 22 at %. Although the source of oxygen was not explained by the authors, it is possible to assume, that an oxidation of the non-stoichiometric TF can occur in the ambient air already after unloading of the samples from the reactor. The authors, in particular, mentioned such a possibility [55, 116].
2. The growth features of SNTF were investigated in [126–129]. It was pointed out, that in case of the use of plasma of the hydrogen-containing compounds (NH_3, H_2) a blocking occurred of the active places of the surfaces (termination), causing a decrease of the deposition rate. The maximal values of the deposition rate were reached, when the nitrogen plasma was used.
3. The authors [127] noted the influence of the residence time parameter (τ, s) in the plasma flow chamber on the rate of deposition

and properties of SNTF, and called the discovered effect "a redeposition." The value of τ was defined as the relation of the known volume of the reactor to the known volume consumption of the plasma-activated gas through the reactor. The effect of redeposition was observed at $\tau = 0.1 - 0.9$ s. According to the authors, the essence of the effect consisted in the fact that at big τ, i.e., at a slow movement of the plasma-activated gas in the reactor, the reaction products could get back to plasma, then they disintegrated and were repeatedly deposited on the growing film. In particular, at an increase of τ in the range of 0.1 – 0.9 s the following changes occurred: the refraction index decreased in the range of 1.95 – 1.72, the concentration of oxygen – 2.5 – 5 at %, the concentration of carbon – 0.5 – 1.5 at %, while the etch rates increased in four times.

4. With the growth of temperature in the range of 100 – 500°C the rate of growth decreased within 0.09 – 0.015 nm/cycle at duration of a cycle of about 17 s; in the same range of temperatures the SNTF characteristics were: RI = 1.63 – 1.96, Si/N = 0.37 – 0.66, the concentration of carbon decreased from 25 down to 2 at %, and the concentration of hydrogen – from 11 down to 5 at %. An increase of duration of a plasma pulse of nitrogen in the range of 1 – 15 s led to a certain decrease of the growth rate within the limits presented above, to variation of the concentration of carbon in the range of 18 – 8 at % and to an increase of correlation Si/N in the range of 0.4 – 0.62. At that, irrespective of the growth conditions, oxygen was discovered in TF with the concentration of about 5 at %, the source of which was recognized as the quartz walls of the reactor. The quality of SNTF is determined by the deposition temperature: if it is over 500°C, the films are close to the stoichiometric by their composition and basic properties. Carbon embedding in SNTF is described in the form of formation of Si–C and C–N bonds [129]. We should point out, that this corresponds to the conclusion of the authors [131] about saturation with carbon of SNTF obtained during a thermally activated ammonolysis of the same reagent. In fact, the

given precursor BTBAS was studied during a decade earlier by the authors of the works quoted above for the purpose of obtaining of SNTF films in LPCVD by reaction with ammonia at the temperature below 600°C. In particular, the authors demonstrated, that in comparison with the traditionally used dichlorosilane, application of the organic precursor, at the acceptable level of the deposition rates of 0.4 – 3.0 nm/min and uniformity of the thickness of coating on the wafers of 200 mm, obviously led to occurrence in the films of carbon (about 7 at %) of the embedded into the matrix material instead of nitrogen, and to a 5 times higher content of hydrogen (up to 15 at %).

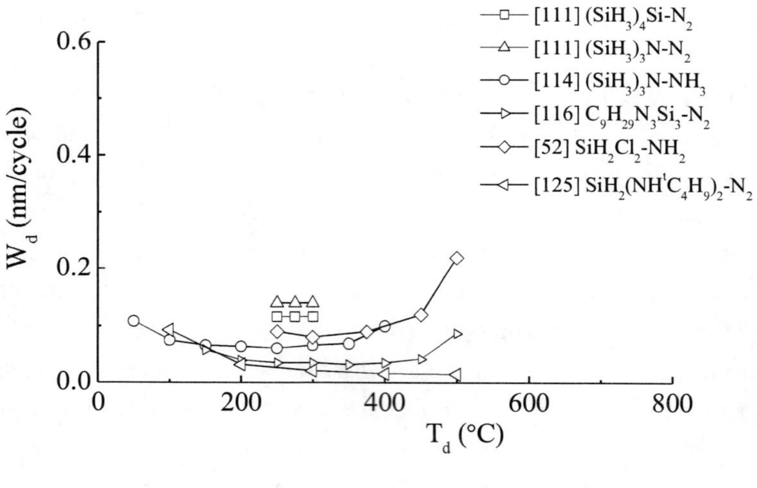

Figure 12. W_d vs T_d dependences for PE-ALD processes in cyclic reactors.

5. Figure 12 shows W_d vs T_d dependences for plasma activated ALD processes with the use of different silicon precursors. One can see that compared to thermally-activated ALD data in Figure 6 the deposition rate values are significantly (in four-fifth times) lower. Regions 3 assumed to reflect "ALD windows" can be proposed in between 200 and 450°C.

6.5. Comparison of ALD Processes with the Other SNTF Deposition Processes

Development of the processes for obtaining the thin films goes alongside with the development of the devices of the micro- and nanoelectronics and MEMS, which use or intend to use, also the materials known under the more general name of SNTF. This, for example, explains that the processes of the thermally activated ALD (further, TA-ALD) were investigated at the turn of the 2000s, while the intensive studies of PE-ALD were begun at the turn of 2010s when the IC technology nodes reached the level of 45 nm and less. Improvement of the technology as a whole demanded a smaller thickness of the thin films of all types. At that, the key parameter of any kind of ALD was the conformality of the deposition. The author selected two directions for comparison of information concerning PE-ALD SNTF: with TA-ALD processes [38] (described in Section 3) and with the processes of a low-temperature plasma-activated deposition in the flow PECVD reactors (described in sections 4,5). The standard for comparison are stoichiometric SNTFs (S-SNTFs) obtained by the industrial method of ammonolysis of dichlorosilane in low pressure reactors.

6.5.1. TA-ALD vs. PE-ALD Films

In both cases, very few experimental data concerning the characteristics are available, the temperature dependences are estimated only in several works. The published results, as a rule, are presented by the separate research groups developing their directions, and the results are not confirmed by the other authors as, for example, it was with obtaining of SNTF in LPCVD in the 1970–1990s. In [38] according to about 10 publications it was established, that the temperature of TA-ALD could be lowered by 200°C and more, compared with the industrial process of ammonolysis of dichlorosilane (~700°C). At that, the author would like to mention the recently published patent application [132] of Applied Materials, Inc., in which TA-ALD, employing $SiCl_4$ and the nitrogen-containing reagents, was announced in the range of 600 – 725°C, i.e., only by 50 – 100°C lower than the object of comparison. At that, the level of concentration of hydrogen was

10 at % for 600°C and it lowered down to 6 at % at 700°C. Unfortunately, the publications considered in [38] and [132)] did not present concrete results about the improvement of conformality for TA-ALD in comparison with the processes of S-SNTF deposition in the LPCVD reactors. Thus, we have enough grounds to consider that the known processes with the thermal activation (sections 2,3) are rather close by characteristics of the processes, composition, and properties. However, the data presented in Table 5 and in [38] concerning the ALD processes point to the abnormally large doses of the precursors used for achievement of an acceptable rate of deposition (in [132] such data are not available at all).

The values of the presented doses of the reagents provided an opportunity for the author to undertake a quantitative estimation of the potential efficiency of TA-ALD processes for industrial production. For estimation of the values and sufficiency of the doses of the reagents at PE-ALD in comparison with TA-ALD, the author accepted the following initial data: density of the stoichiometric Si_3N_4 — 3.17 g/cm^3 (in fact, it is less), the molecular mass of Si_3N_4 – 140, substrate diameter of 200 mm (area –314 cm^2), averaged growth rate – 0.2 nm/cycle (0.2×10^{-7} cm/cycle). According to these data, estimation of the volume of SNTF formed on the substrate per one cycle of the process equals $\sim7\times10^{-8}$ cm^3/cycle, or 2×10^{-5} g/cycle, or $\sim9\times10^{16}$ molecule/cycle. Taking into account the real deviations from the accepted values towards reduction of the deposition rate down to one order, the real deviation of TF density to the smaller value, etc., it is possible to assume, that the value of $\sim10^{17}$ molecule/cycle for SNTF formed per a cycle corresponds well to the published experimental data.

For estimation of the characteristic of a pulse of the precursor following the known Mendeleev–Clapeyron equation we will accept the volume of an individual reactor as ~5 dm^3, the gas constant – 62.3 Torr×dm^3×mol^{-1}×K^{-1}, deposition temperature – 400°C (700 K), pressure in the reactor – 100 Torr (according to data in [38], the used pressure was up to 500 Torrs). Estimation of the number of molecules in a pulse for the above values gives a value of $\sim7\times10^{21}$ molecules. For a dose of the precursor of one L, the estimation of the number of molecules equals $\sim7\times10^{13}$ molecules (which means that in the pulse, calculated above the dose will be $\sim10^{8}$ L). Proceeding from the estimations presented above, the precursor dose, necessary for the formation of one monolayer of SNTF is at the level of ~10^4 L, which by 4 or 5 orders less, than the data presented in the publications on the experimental doses of the precursors at TA-ALD. The presented estimation allows us to conclude that the efficiency of the use of the precursors at TA-ALD (the fraction of the precursor transferred into TF from the quantity of the precursor injected into the reactor) equals to only hundredth parts of percent, which, no doubt, is insufficient for production. For comparison: according to the author's data, the efficiency of the industrial process of ammonolysis of dichlorosilane in LPCVD is about 5% [2)], and for PECVD – about an order higher. It is necessary to point out that for PE-ALD the doses, necessary for the precursors, turn out to be by several orders less, which testifies to the higher efficiency of such processes.

Comparison of TA-ALD and PE-ALD conformality data are presented in Table 6. One can see that achieved results are quite poor even for low aspect ratio values.

Table 6. SNTF conformality data for TA-ALD and PE-ALD processes

Reactants	T_d, °C	Structure type	H (μm)	G (μm)	AR	Conformality (%)	Ref. ##
Si_3Cl_8-NH_3	510	Inclined	8.6	0.14	60	83	[57]
SiH_4-N_2*	250–400	Inclined	0.17	0.064	3.8	50	[110]
$(SiH_3)_4Si$-N_2*	275	Vertical	1.4	0.35	4	35**	[111]
$C_9H_{29}N_3Si_3$-N_2*	300	Inclined	0.4	0.075	5.3	80 (73**)	[116]
$SiH_2[N(C_2H_5)_2]_2$-N_2*	300	Inclined	0.4	0.075	5.3	80 (57**)	
SiH_2Cl_2-NH_3*	500	Spacer	0.05	-	-	100	[121]
Si_2Cl_6-NH_3*	400	Inclined	0.4	0.13	4	100	[123]
$(SiH_3)_3N$-H_2/N_2*	300–400	Spacer	0.075	-	-	85	[112, 113]

Notes. *Plasma activations; ** conformality in 1/2 of the structure depth.

6.5.2. PECVD vs. PE-ALD

Since it is considered, that PE-ALD processes give a chance to lower the temperature of "the ALD window" by several hundreds of degrees, this temperature interval gets into the interval of the temperatures, characteristic for the PECVD and HDP-CVD processes. The basic difference of PE-ALD from the PECVD is a very low, by orders less, rate of deposition in the comparable temperature intervals. However, the general trend for reduction of the thickness of the films for modern technologies down to the level of several tens of nanometers smoothes this drawback. At the same time, the conformality of certain processes of PE-ALD in comparison with PECVD is essentially better than it was confirmed by the authors [112, 113, 116, 121–123]. Besides, by composition, such films come closer to the stoichiometric ones. Now, the PE-ALD processes are far from being recognized as mature and still developing. The author would like to note the following basic trends in such development. The first one is the search activity for the precursors, not containing chlorine and carbon. The second one is the use of the reactors of "direct plasma," that is, with a capacitive excitation of the plasma, the simplest reactors, which are identical to the industrial reactors of PECVD. Here, it is necessary to point out, that an approximate level of the necessary doses at PE-ALD is realized at rather long durations of the pulses of the reagents, reaching tens of seconds (see Table 5), which is quite comparable with the duration of short production PECVD processes. It is possible to assume, that the well-known (and cheaper) plasma CCR can turn out to be a good variant for the realization of the PE-ALD processes with reference to the low-temperature SNTF of satisfactory quality. As a topical problem for increasing the productivity of the PE-ALD processes, the author would mention a search for the ways of surface processing for increasing the deposition rate. Such works, in particular, allow us to expect the use of the activated nitrogen [124] or hydrogen [53]. However, processing of the internal surfaces of the complex reliefs of the devices will demand the use of the intensive flows of the activated particles, i.e., an increase of the exciting energy, capable to cause unacceptable infringements in the devices.

6.6. Conclusion to Section 6

The plasma-activated ALD allows us to obtain SNTF at temperatures below 500°C from the precursors of an inorganic nature (silanes, chlorine derivatives of silanes, amines). Characteristics of the coatings meet the requirements of IC manufacture, ensuring acceptable rates of growth and low wet etch rates in the liquid etchants based on hydrofluoric acid. The conformality of the growth of coatings at PE-ALD considerably exceeds that of PECVD. The author would like to mention the following trends in the development of the PE-ALD processes. The first trend is the search activity concerning the precursors: the precursors of organic nature are promising, allowing us to obtain SNTF with the characteristics acceptable for the modern micro- and nanoelectronic technologies. At the same time, it is necessary to point out, that despite the long-term and repeated attempts to introduce the organic precursors into TF production, it is possible to consider them failed in the cases when acceptable precursors of the inorganic nature are available. The second trend is the use of CCR – the simplest and, actually, identical to the industrial classical reactors of PECVD. The third trend is the search for ways for processing the surface for an increase in the rate of deposition.

7. Hot-Wire CVD in Flow-Type Processes

7.1. Brief Description of Hot-Wire CVD Processes

In previous sections, the author analyzed the state of technologies for obtaining silicon nitride thin films (SNTF) in the flow and cycle types of reactors for IC and MEMS technologies. It has been shown that the deposition temperature can vary from about 800°C (LPCVD, thermal activation) to less than 400°C (PECVD, plasma activation). In the case of activated low-temperature CVD processes, the non-stoichiometric materials are formed, designated by the author as "silicon-nitrogen-hydrogen containing thin films" (SiNH-TF). The main undesirable SiNH-TF component for the technology is hydrogen with a concentration up to 40 at %, which deteriorates the technologically important properties of the films (density, wet etch, etc.). Use of HDP-CVD ensures many times lower and acceptable for the technology level of content of hydrogen, essentially higher densities of the films and lower rates of dissolution. However, this method historically was dedicated to semiconductor substrates with complicated device relief. Thus, the need for a low-temperature TF deposition method without plasma activation (i.e., with thermal activation) suitable for diversified forms of the substrates became obvious.

In the late 1980s, a method was proposed for obtaining SNTF, known under the names Hot-Wire CVD, Hot-Filament CVD, or Catalytic CVD [133]. The essence of the method consisted in the deposition of a thin film in regular flow vacuum reactors to the substrate heated up to a low (usually less than 400°) temperature by a realization of a chemical reaction of monosilane SiH_4 with a mix of nitrogen and hydrazine, N_2H_4. At that, unlike in the classical methods of CVD, a gas flow is supplied to the substrate through a tungsten wire heated up to temperature T_w = 1200 – 1400°C and situated at a distance of several centimeters from the substrate. As a result, SNTFs were formed on the substrates with a significant rate of growth. At that, the quality of the films was close to that of the plasma methods for their obtaining (that is, to SiNH-TF) and acceptable to various instrument applications, demanding very low temperatures. Due to rather high working pressure (about dozens-hundreds of Pascals) the method appeared to be rather simple in the hardware respect, not demanding complex (expensive) vacuum pumping systems, and, hence, relatively inexpensive. The given method for reception of SNTF in the flow reactors hereinafter called HW-CVD attracted a lot of attention of the researchers and is topical even now.

The reactors for HW-CVD are the flow reactors functioning in the mode of the dynamic forevacuum. Mainly, the publications describe regular CVD flow chambers from stainless steel, intended for one or several substrates. Such chambers can be modernized easily for HW-CVD by the introduction of a module with an activating wire. An example of a rather typical simplified design of a reactor for HW-CVD is presented in Figure 13(a). The author would like to single out two publications, in which the questions of equipment for HW-CVD are considered. In [134] the results of industrial commissioning of the commercial installations for the reception of different purpose thin films (including – SiNH-TF) are analyzed, and also a comparison is done of the specific features of deposition of SNTF for the HW-CVD and PECVD processes. Certainly, of interest are the industrial variants of the deposition systems for large rectangular-form objects, and also rolled materials, realized on the "roll-to-roll" principle. Also of interest is the author's analysis of a set of questions concerning the use of the activating wires.

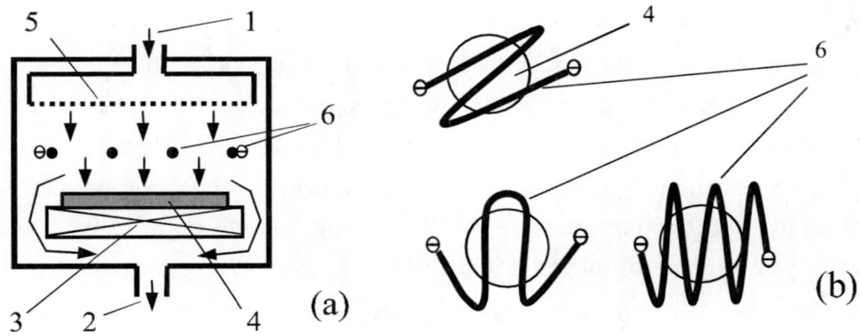

Figure 13. Simplified sketch of HW-CVD reactor (a) and some examples of hot wire configuration with respect to the round substrate position. Definitions: 1 – reactant "in": 2 – reaction by-products "out"; 3 – heater; 4 – substrate; 5 – gas mixture assembly or showerhead; 6 – hot wire with edge connectors to power supply. Arrows indicate gas flow direction through the hot wire.

In [135] the questions are considered of the influence of the geometry of the arrangement of a wire in a reactor on the substrate temperature (without its additional heating), the rate of deposition, and its uniformity on the substrate area. In particular, at the value of the gap of about 5 cm, the wire-substrate was heated in the course of deposition up to 190 – 250°C depending on the length of the wire and the way of its arrangement over the substrate, Figure 13(b). However, at an approximately double increase of the gap the temperature decreased 3.5 times. Thus, heating of the objects up to approximately 200°C occurs utilizing the activating wire itself, while for the use of the lower temperatures of the substrates their compulsory cooling is necessary.

7.2. Brief Review on HW-CVD Processes and SiNH-TF Properties

The actual data about the HW-CVD processes and SiNH-TF properties are presented in Table 7. In one of the first works on Cat-CVD [133], the heating of a hydrazine and nitrogen mix was carried out by a wire with a composition of 2% Th–W, the diameter of 0.45 mm, and with a total area within 15 – 42 cm^2. The distance between the substrates and the heated wire was 3 – 4 cm. The composition of the films (in terms of Si/N ratio) varied within a wide range depending on the correlation of the components: a reduction of the partial pressure of the nitrogen-containing substances caused the enrichment of thin film with silicon. In the FTIR spectra the peaks of the bond absorption were discovered with the maxima for Si–N bonds at 830 cm^{-1}, N–H bonds at 3320 cm^{-1}, and Si–H bonds at 2180 cm^{-1}. The resistance of SiNH-TF was 10^{13} – 10^{15} $\Omega\times$cm, which roughly corresponded to the films obtained by PECVD. The conformality of the coatings for a rectangular form trench with the width of 10 μm and aspect ratio of ~5 was unsatisfactory because that was just about 50%.

In [136], a tungsten wire of 0.25 mm was used with the value of the substrate gap with a diameter of 50 mm and about 4 cm thickness. The concentration of hydrogen decreased with the rise of the temperature of the substrates and the reduction of the ratio of ammonia to monosilane. In the films, the presence of oxygen was found in the areas closer to the surface and the boundaries with the other films. As a source of silicon, the authors [137–139] investigated disilane Si_2H_6. Wet etch rate in BOE was within the range of 2.5 – 7.9 nm/min for SiNH-TF with a refraction index within the range of 2.08 – 2.04. At that, in the comparison sample of SiNH-TF received during PECVD with the refractive index of 1.88 and concentration of hydrogen about 12 at %, wet etch rate was many times higher and equal to 51.8 nm/min.

In [140–142], research was done on the installation based on monosilane and ammonia, including an algorism [133], in which hydrazine was investigated. Estimations of the concentration of oxygen, the resistance of SiNH-TF, and conformality of their deposition on the same structures gave similar results to [133]. The authors [143] studied two so-called "inert" types of SiNH-TF (which were obtained at 430°C at a small excess of ammonia) and the "porous" ones (obtained at 340°C at a large excess of ammonia).

Table 7. Summary of research on SiNH-TF by HW-CVD

Ref. ##	**Reactants**	**T_w (°C)/ T_d (°C)/ P_d (Pa)**	**W_d (nm/min)**	**Si/N**	**RI**	**[H], at.%**
[133]	SiH_4, N_2, N_2H_4	1200–1390/ 230–380/ 7–1000	20–90	0.6–3.3	~1.8	-
[136]	SiH_4, NH_3, H_2	1747/ 393–636/ 66.5	50–250	0.8–0.9	1.86–1.99	2.4–34
[137-139]	Si_2H_6, NH_3, H_2, Ar	1500–1850/ 245–370/ 66.5	61–171	0.88	1,86–2.42	6–17.7
[140, 141]	SiH_4, NH_3	1700–1800/ 200–400/ 0.13–5.3	4–8	0.75–2.5	1.9–2.5	-
[142]	SiH_4, NH_3	1800–1900/ 300/ 1.33	-	0.75	2,0	-
[143]	SiH_4, NH_3	1900/ 300–500/ 1–10	3–21	0.71–0.95	1.88–2.3	10–15
[144]	SiH_4, NH_3	2100/ 450/ 20	180–240	1.33–2.5	1.95–2.55	9–16
[145]	SiH_4(He), NH_3	1800/ 300/ 2.66–13.3	-	0.99–2.5	1.8–2.5	9.3–16
[146]	SiH_4, NH_3, H_2	1650/ 100/ 8	-	0.77–0.8	-	-

The values of the dielectric constant for these types appeared to be at the level of 6.3 and 7.9, accordingly, the value of the fixed charge for porous films was 5.8×10^{16} cm^{-3}, however, such films turned out to be inclined to oxidation after the process in the air. In [145] a wire was used with a thickness of 0.5 mm, length of 12 cm, and a gap between the wire and the substrate of 5 cm. The energy of the reaction activation was estimated as 31 kcal/mol. In [146] the correlation of certain electric properties and parameters of the HW-CVD process were investigated. For example, the breakdown voltage was $(4–8)\times10^6$ V/cm, the specific resistance equaled to $1.0\times10^{11} – 1.5\times10^{12}$ Ω×cm, and the value of the dielectric constant – 4.8 – 6.8. In [147,148] the data are presented concerning the surface roughness and mechanical stresses in SiNH-TF, received with super high rates of deposition, up to 430 nm/min. For the compositions, close to the stoichiometric ones, the roughness was about 1 nm. The authors discovered that for the composition characterized as SiHx, within the interval of $1.2 < x < 1.6$ the stresses in the films had a tensile character, at that, with an increase of the concentration of nitrogen the stresses increased. For the samples with Si/N = 0.8, the value of the stresses was 16 MPa, which was essentially less, than for PECVD SiNH-TF (section 4).

Generalizing the results, we can conclude the following. The hardware realization of the HW-CVD processes described in the literature was organized approximately in the same way, at that, the key parameter was the gap between the activating hot wire and the substrate, which was about 4 cm. We should point out that this was many times more than the gaps between the electrodes in PECVD installations. A red-hot wire heats a sample to temperatures over 200°C, which means that achievement of higher temperatures of the substrates requires either optimization of the size of the gap or a special additional heater, while achievement of lower temperatures requires a special cooling system for the substrates. In general, the results demonstrate a possibility of obtaining SiNH-TF at HW-CVD with a rather wide range of Si/N correlations, including the correlations, close to the stoichiometric one of 0.75.

7.3. Brief Review on HW-CVD SiNH-TF Reaction Mechanism

There is no sufficient information on the results of the research of the mechanism of growth of SiNH-TF at HW-CVD. In [148] the efficiency of the utilization (use) of the reagents was estimated at a very high level – up to 98% for SiH_4, about 50% for NH_3 (for the case of small flows of gases), and about 70% for SiH_4 for the case of high speeds of the flows. In [149] the products of decomposition of monosilane on a heated wire were investigated, and the basic ones were established as Si, SiH_3, Si_2H_6. In [150] certain data on the kinetics of the growth of SiNH-TF and research of the gas phase in the course of the growth were generalized. The authors [151] presented the experimental data concerning the formation of the products of the reaction of monosilane with ammonia with a correlation of 1:1–200 on a hot wire with the use of ND3 isotope. It was established, that the composition of the products was determined by a ratio of the reagents in gas flow. At a high content of ammonia, the reaction route was described through the formation of NH_2 and aminosilanes. In case of an insufficient excess of ammonia, the processes of decomposition of monosilane occur with the formation of SiH_3 and then of Si_2H_6 and Si_3H_8. Indirectly this means a possibility of enrichment of a film by silicon. Thus, for obtaining the compositions close to stoichiometry it is expedient to use essential excess of ammonia.

7.4. Comparison of HW-CVD and PECVD Processes for SiNH-TF

As a result of analysis of a number of the published experimental data on SiNH-TF at HW-CVD, it is possible to note their small volume and absence of any generalizing comparative works as it takes place for the films received by the other methods (see above sections). An essential feature of the HW-CVD processes is the use by the majority of the authors of the most simple and industrial reagents – monosilane and ammonia in the reactors, very close by their design. As a whole, the properties of the films are close to the low-temperature PECVD, however, the concentration of hydrogen in HW-CVD SiNH-TF on average does not exceed 15 at %, which is essentially less, than for PECVD (up to 40 at %), and approximately corresponds to the level of the processes of HDP-CVD. A comparison of the generalized characteristics of PECVD and HW-CVD is presented [152] in

Table 8 based on summarized information from references [139, 147, 148]. Some obvious advantages of HW-CVD can be seen.

First of all, this refers to the forms and sizes of the substrates, opportunities for application of the processes for the flexible substrates, the influence of the active particles on the instruments, and also the efficiency of the use of the reagents. A quantitative comparison of the HW-CVD processes with the other processes, analyzed in sections above, is complicated. By such a comparison the author means an estimation of the effective constant of the deposition rate of the process for HW-CVD (k_{eff}), which is the ratio of the rate of growth Wd to the concentration of a silicon-containing reagent in the gas phase [C_{Si}]: k_{eff}~W/[C_{Si}]. For the slow CVD processes it was estimated at the level of 0.05 cm/s, and for the fast ones, with plasma excitation – at the level of 10 cm/s [2].

The problems with estimation for HW-CVD are explained by two reasons. First, in the literature, there are no accurate data about the sizes of the reactors. Secondly, the basic difference with the realization of the HW-CVD method is heating of the activating wire up to a very high temperature and heating by such wire of a substrate up to low temperatures, depending on the size of the wire-substrate gap. The author attempted to rather approximately estimate the values of k_{eff} by a set of data about the processes presented in [136, 141]. By using the methods for estimation of the processes in [2], the author estimated the value of k_{eff} at the level of ~70 cm/s according to the data [136] and ~240 cm/s according to the data [141], which was almost 10 times more than the values received for the plasma-activated processes. Notwithstanding its conditional character, the given estimation is important, first of all, because it unequivocally points to the essentially higher k_{eff}, at HW-CVD. This, on one hand, is correlated directly with the efficiency of the use of the reagents, and on the other hand, points directly to the principal non-conformality of SiNH-TF deposition in this process, because at CVD the conformality is qualitatively connected with k_{eff} [2]. This is confirmed by the qualitative experimental data [133, 140, 141].

Table 8. Basic differences between Cat-CVD and PECVD processes

Item #	HW-CVD	PECVD with the capacityively coupled plasma
1	A reaction is initiated by a catalytic cracking of molecules on a hot metal wire	A reaction is initiated by fast electrons in a radio-frequency field
2	The gas phase of the process does not contain ions	The gas phase of the process includes radicals, neutral particles, electrons and ions
3	The substrate does not participate in the process of the thermal disintegration of the initial reagents	The substrate is a part of the process and should have an equipotential surface for creation of a homogeneous plasma
4	Formation of radicals occurs during collision of the molecules of the reagents with a hot wire, in essence, by the reaction of the gas with the wall	Dot collisions are necessary for formation of the radicals in the three-dimensional space
5	Undesirable microparticles can be formed only at a high pressure	Negatively charged particles can be captured in the area of the positive plasma and become a source for formation of the undesirable microparticles
6	The growing layers are not subjected to the ionic bombardment	The growing thin films are subjected to the ionic bombardment
7	The substrate temperature is influenced by the wire heating, optionally, a special heater/cooler can be used	The substrate temperature is set by a ~~a~~ special heater
8	Cleaning and maintenance service schedules are not established yet and are a result of the experimental research.	The cleaning procedures and maintenance service schedules, which ensure the quality of TF, are known
9	The reactors are scaled easily for any sizes and forms of the substrates	Scaling of the reactors for the other sizes and forms of the substrates demand essential efforts
10	The work with the flexible substrates and rolled materials is possible	The work with the flexible substrates is complicated
11	Efficiency of the use of the reagents is at the level of up to 70–90% on monosilane and about 50% on ammonia	Efficiency of the use of the reagents is at the level of 15–20%

7.5. Directions for HW-CVD Processes Development

The obvious directions for the development are connected with the advantages of the HW-CVD processes presented in Table 8. Among them it is possible to name the development of equipment for obtaining coatings on the substrates of large sizes and rolled materials, improvement of the systems of arrangement, fastening and replacement of the activating wires, improvement of the reproducibility and uniformity of the deposition of the films. Apparently, from the point of view of the known tasks of the microelectronic technologies, first of all, obtaining the conformal coatings on complex IC reliefs, application of this method in microelectronics is unlikely. The possibility of obtaining coatings at low temperatures with a low level of mechanical stresses can be of interest for MEMS technologies, however, it will demand special equipment for deposition. At that, technologically such equipment does not demand the use of any unique materials or devices, i.e., can well be designed and made for specific aims. It is also interesting to point out, that the very high efficiency of the use of the initial reagents reduces the problems of recycling of the exhausts of such technological installations.

7.6. Conclusion to Section 7

The processes of hot-wire deposition of SNTF are in many respects unique because they allow us to receive compositions, close to the stoichiometric ones, without a plasma activation of the gas mixes and plasma influence on the substrates with the devices' structures. The obvious advantages of the HW-CVD processes are a possibility of obtaining thin films with a wide range of compositions, including close to the stoichiometric ones, and with a low content of hydrogen, a possibility of the use of low temperatures of deposition for substrates of various forms and the sizes, including flexible substrates, and high efficiency of the use of the initial monosilane and ammonia. A drawback of the HW-CVD processes is a bad conformality of deposition of the thin films. In comparison with the low-temperature PECVD processes, now the HW-CVD technologies are, more likely, at the stage of their development.

8. SNTF Composition and Properties

8.1. Relationships between CVD Process Parameters and Thin Film Quality

For practical use, thin films should demonstrate some important characteristics, which can be considered necessary for device technology. Some properties of S-SNTF and SiNH-TF films have been presented above in table format. These data can be used directly, but they do not show any correlations with CVD process characteristics. In [2] a generalized approach was formulated to the description and analysis of the processes of obtaining TF of various types with the participation of the gas phase reactions. The approach consisted of a presentation of four basic dependencies: “output” characteristics of the process of deposition on the “input” parameters of the processes (temperature, pressure, concentration of the reagents, etc.). Those major for the TF technology characteristics of the processes are the following:

1. Rate of deposition (rate of growth) of TF;
2. Composition of TF;
3. Physical and chemical properties of TF;
4. Electrical and electrophysical properties of TF.

The aim of this Section is a systematization of the data for the first three of the above characteristics concerning various methods of deposition and types of SNTF considered above.

At the depth of search of several decades the following methods were considered, which in the text below are designated, accordingly, by numbers 1 – 6:

- Method M1 – in the flow reactors during the chemical vapor deposition from the gas phase with the thermal activation either at LPCVD in hot-wall tubular reactors or single substrate reactors;

- Method M2 – in the cyclic reactors during the thermally activated atomic-layer deposition (TA-ALD);
- Method M3 – in the reactors of the flow type during the plasma-enhanced chemical deposition (PECVD);
- Method M4 – in the flow type reactors during the plasma-enhanced chemical deposition with high-density plasma activation (HDP-CVD);
- Method M5 – in the cyclic reactors during the plasma activation of ALD (PE-ALD);
- Method M6 – in the flow reactors with the so-called "hot-wire deposition" (HW-CVD, i.e., during a thermal reaction of the hydrides of silicon and nitrogen on a heated tungsten thread.

The analysis below covers the overwhelming majority of the technological applications of SNTF and is necessary for the formation of a complete comparative view of the opportunities presented by various methods of SNTF deposition. The author would like to underline, that in the given work the electrical and electrophysical properties of SNTF are not considered (the fourth type of SNTF characteristic). The reason for it consists in an essentially smaller volume of the publications concerning the limited methods for obtaining and specific use of TF, going beyond the framework of the given review.

In the present work, as the basic "input" parameter for the generalization of the resulting information on the deposition processes, the temperature of deposition of SNTF (T_d) was selected. High-temperature LPCVD methods (T_d > ~700°C) allow us to obtain S-SNTF. At lower temperatures, depending on the obtaining method, the more or less nonstoichiometric SiNH-TFs are formed. However, also in the latter case, by manipulations with the composition of the gas mixes and conditions of deposition, we can get SNTF, formally close by their composition to the stoichiometric ones. This occurs in case, if we take into account the composition of the characteristic of only the ratio of Si/N = 0.75, however, as it is shown below, the correlation of Si/N is not an exhaustive characteristic of the composition of SiNH-TF. During our analysis in the present work, we considered the generalized values of the characteristics and parameters, the initial data of which can be found in the related sections above.

8.2. Temperature Dependences of SNTF Deposition Rates

Figure 14 presents the shaded in different ways the approximate values of the rates of deposition (W_d) for methods 1 – 6 in case of the temperature intervals typical for them. These regions are rather referenced ones because within the shaded areas there are some variations of the deposition rates.

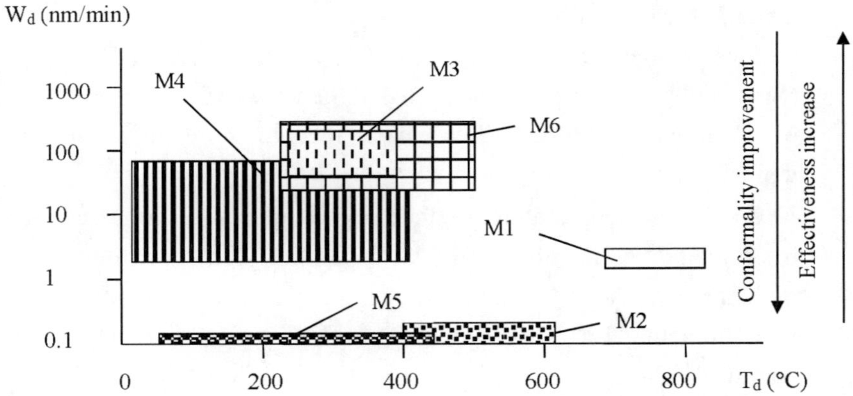

Figure 14. Exemplary ranges of deposition rate values for temperature intervals in which the deposition of SNTF is carried out by the methods of LPCVD – M1 [12], TA-ALD – M2 [38], PECVD – M3 [99], HDP-CVD – M4 [153], PE-ALD – M5 [154], and HW-CVD – M6 [152].

However, the convenience of such a presentation consists in the fact that through the shaded areas it is possible to present visually the possibilities of the methods for obtaining of SNTF, and also to pass to the generalizations concerning the composition and the basic properties of the films. Further in the work, the chosen types of the shaded areas for the considered methods of deposition M1 – M6 are preserved.

To the data presented in Figure 14 it is possible to add the following comments:

- the values of the rates of deposition for methods M1 – M6 differ more than by four orders;
- methods TA-ALD (M2) and PE-ALD (M5), realized, as a rule, in the cyclic reactors of individual type (that is, on one substrate), allow us to obtain films with extremely small rates of deposition, less than 0.1 nm/min. In this case, the author believes, it is admissible to use nm/min dimensionality instead of nm/cycle dimensionality typical for the ALD processes, because the duration of the cycles in both studied versions of ALD was equal to a minute and over. From the presented data it is visible that the ALD methods described in literature now are rather not suitable for use and demand further research;
- the plasma-activated methods of PECVD (M3) and HDP-CVD (M4) and the hot-wire method (M6) in the flow reactors, more often of an individual type, allow us to receive very high rates of deposition;
- LPCVD high-temperature method (M1) allows us to reach the intermediate values of the rates of deposition. However, due to its realization in the batch flow tubular reactors with a simultaneous load of about 100 substrates, high efficiency of the process is achieved. The tubular high-temperature reactors with hot walls are applied in large quantities in the manufacture of IC and MEMS for obtaining S-SNTF during the use of SiH_2Cl_2. A rare, but interesting for deposition at high temperatures from SiH_4, is the flow, hot-wall single substrate reactor. An advantage of such a reactor is essentially smaller cumulative thermal influences in comparison with the traditional LPCVD method in the tubular reactors. This is due to higher, up to 200 nm/min, rates of growth of the films because of the use by more than 2 – 2.5 orders higher working pressures (the rate of deposition, as a rule, is directly proportional to the pressure in the reactor). An advantage of the use of monosilane is also a lower defect density of the films due to the absence of chlorine in the initial molecule. This eliminates the problem of the formation of microparticles of the ammonium chloride (NH_4Cl) and other

chlorine-containing compositions in the gas phase and on the walls of the vacuum systems, pronounced for the silicon chlorides. The author would like to point out one more potential advantage. It is expected, that in such reactors due to variation of the correlation of the components in the gas phase it is possible to obtain high-temperature films with a homogeneous composition and silicon enriched content (SE-SNTF). According to data [17–19], such films are of interest for use in MEMS technologies. As it was discussed in [12], obtaining SE-SNTF in the tubular flow reactors during LPCVD is impossible because of a big variation of the concentration of the components along the length of the working zone, which is eliminated in principle in the discussed individual reactor. Obtaining very thin S-SNTF films for a rather narrow application is also possible by the method of a rapid thermal deposition (Rapid Thermal CVD, RT-CVD) from $SiH_4 – NH_3$ mixes [155]. Within the range of 700 – 850°C the rates of growth in the reactors of individual type are equal to 50 – 300 nm/min, at that, the pressure in the reactor by an order exceeds the pressure in the "classical" tubular LPCVD reactors.

Giving brief characteristics to the processes of SNTF deposition, the author would like to point out, that for methods M1, M2, M5 the processes were described as proceeding following multi-route schemes with the limiting heterogeneous steps. In such processes, the substrate surface is an active component of the reaction. At the same time, as a rule, the plasma-activated processes proceed in the gas phase without the participation or with minimal participation of the surface of the substrates. As a certain exception from this, we can consider some PE-ALD processes, for which the plasma activation of one of the reaction components can increase the contribution of the surface reactions.

For characterization and comparison of the kinetic features of the deposition processes, in [2] the author introduced a universal characteristic of the processes of deposition of TF (including the one applicable and for processes of obtaining SNTF), named as the effective constant of the rate of reaction k_{eff}. High rates of deposition of the plasma- and catalyst-activated processes correspond to the greater values of k_{eff}. At that, it is necessary to point out, that an increase of W_d (i.e., k_{eff}) also directly correlates with the increase of the efficiency of the use of the reactants, which is shown by an arrow in the right part of Figure 14. However, an increase of W_d (i.e., k_{eff}) has also a negative effect – it leads to a deterioration of the conformality of SNTF deposition on the developed reliefs of the devices. In the same Figure 14, an arrow points to the opposite direction with the arrow of the efficiency of the use of the reagents. Thus, application of the high-speed processes is expedient for the substrates without relief or with a minimally developed relief, while the low-speed processes allow us to receive essentially more conformal coatings [2]. An exception is the combined process of deposition with CVD and the simultaneous etching of TF, specially developed for obtaining the conformal coatings on the high-aspect structures [153].

8.3. Temperature Dependencies of SNTF Composition

As the basic characteristics of SNTF composition, the author selected the Si/N ratio and the concentration of hydrogen in a film. The latter characteristic is especially topical for the low-temperature deposition methods. We should point out, that for a definition of Si/N relation the authors of the publications used various direct methods for the definition of silicon and nitrogen (including the one with profiling of the elements by the thickness of the films): Auger spectroscopy (AES); x-ray photoelectronic spectroscopy (XPS), secondary ion mass spectrometry (SIMS); spectroscopy of the Rutherford backscattering (RBS), the method for detection of the recoil atoms (ERDA), etc., and for hydrogen definition – SIMS, RBS, ERDA, the methods of nuclear reactions (NMR), and also the widespread preliminary calibrated infra-red spectroscopy with Fourier transform, allowing us to record the concentrations of Si-H and N-H bonds in SNTF. Data in Figure 15(a), a point to a practically stoichiometric composition of the high-temperature films (S-SNTF, Si/N = 0.75), at only several atomic percent of the content of hydrogen, Figure 15(b) [156].

However, for the low-temperature plasma-activated SiNH-TF a very wide range of Si/N correlation with a certain general trend to enrichment by silicon is possible. Under certain conditions of the deposition processes (basically, during the selection of the correlations of the silicon and nitric components of the reactions) and for SiNH-TF it appears possible to reach a formal stoichiometric correlation of Si/N = 0.75. However, it is impossible to consider such films as stoichiometric ones, because of a considerable presence of hydrogen in them, Figure 15(b), many times exceeding the content of hydrogen in the high-temperature S-SNTF (including those, received in the individual reactors [71, 155]). Out of the low-temperature ones, the comparative concentrations of hydrogen in SiNH-TF were observed for the films received by method M4 (HDP-CVD), where SiH_4 – N_2 mixes were usually applied. The maximal values of the content of hydrogen (up to 40 at %) were recorded for the PECVD processes with the use of SiH_4 – NH_3 mixes. As a whole, the data presented in Figure 15(b) correspond to the trend of a decrease of the concentration of hydrogen in SNTF in case of a rise of the temperature for obtaining the films [5]. The author would like to underline, that in IC and MEMS technologies the thin films are often subjected to thermal annealing at high temperatures. In this case, it is necessary to take into consideration information from the sources [83, 88] about the removal of hydrogen from the films at temperatures approximately over 500 – 600°C, which leads to a variation of the composition and structure of the films. According to the author, the problem of the presence of hydrogen in SNTF is of interest for a separate publication. As far as the presence of the other elements in SNTF is concerned, for example, chlorine, oxygen, carbon, here the following comments are necessary:

- in the high-temperature films, the presence of the impurity elements (residual elements from the initial reagents) is recorded at the level below the limit of detection of the methods of the analysis;

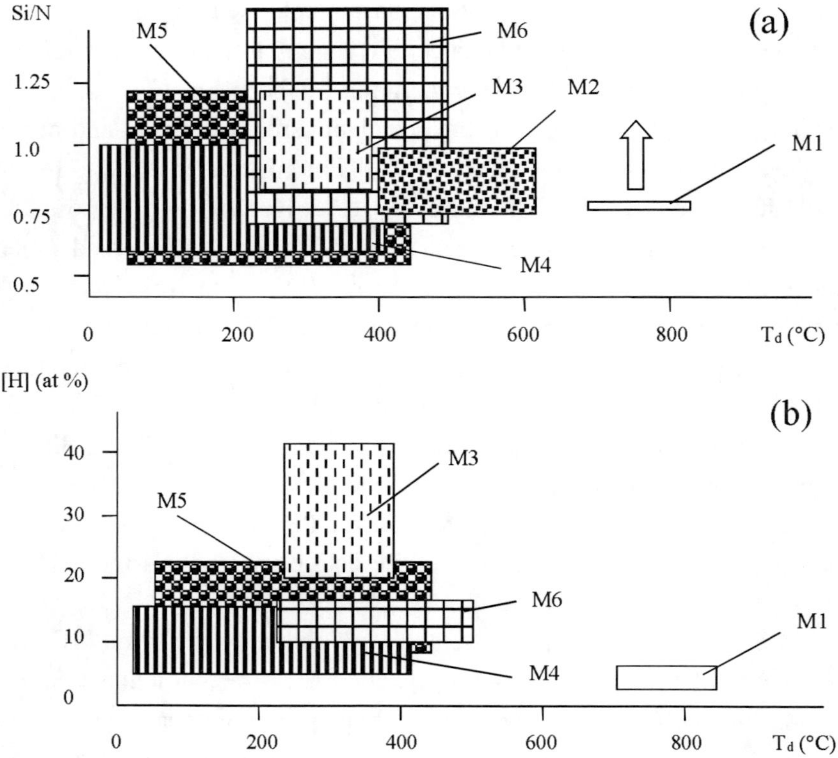

Figure 15. Exemplary ranges of the Si/N ratio (a), and total hydrogen concentration (b), characterizing the SNTF compositions obtained by methods M1 – M6 with the corresponding approximate temperature intervals.

- in the low-temperature films obtained by ALD methods from the chlorine-derivative of monosilane [12, 154], the effects were observed of an abnormal high-speed dissolution of coatings and their oxidation in the air after their extraction from the reaction chambers. According to the information of the work [5], during mixing of the chlorides of silicon and ammonia at low temperatures, the formation was observed of the soluble in water powder-like products: silicon diimide $Si(NH)_2$, hexachlorodisilazane $Si_2(NH)Cl_6$, etc. So, the composition and the structure of thus obtained coatings can be more complex than the regular SiNH-TF because of the presence of chlorine and the hydrogen-containing groups oxidizing easily in the air. However, according to [5], these products can transform into silicon nitride at an essential rise of temperature of the process or under TF annealing. The author would also like to comment on the active involvement by many authors of the values of the refraction index RI for characterization of SNTF composition. The methods for the definition of RI are simple, hardware perfect, accurate, and, as a consequence, widespread in the IC and MEMS industry for control of TF compositions. However, regarding SNTF it is necessary to point out the following. The refraction index for S-SNTF is usually accepted at the level of 2.0 ± 0.02, but in case of the presence of the hydrogen and oxygen impurities in SNTF, RI essentially decreases (down to 1.7 – 1.8), and in case of increase of Si/N correlation it raises considerably (for example, at Si/N = 1.42 the value of RI = 2.5 [144]). Thus, during the research of NTF with the composition varying in a wide range the interpretation of the results through measuring of RI can lead to erroneous conclusions.

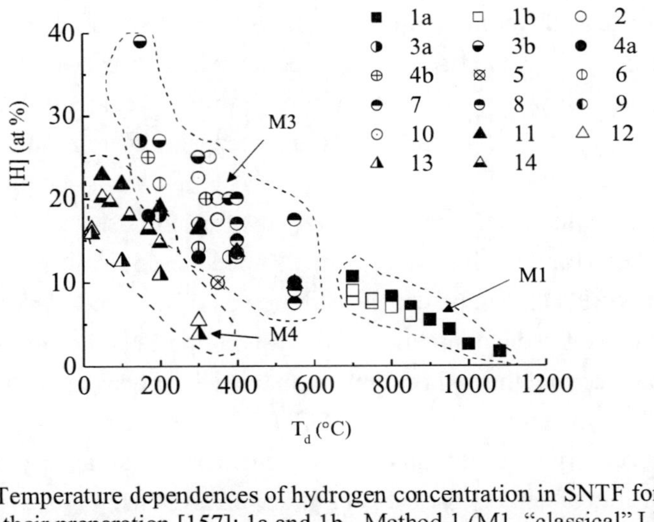

Figure 16. Temperature dependences of hydrogen concentration in SNTF for various methods of their preparation [157]: 1a and 1b - Method 1 (M1, "classical" LPCVD and "fast" thermal RTP, respectively); 2 – 10 - Method 3 (M3); 11 – 14 - Method 4 (M4). The compositions of the mixtures according to the method of 3: 2, 3a, 6 - SiH_4-NH_3; 3b, 5 - SiH_4-N_2; 4a - SiH_4-H_2-Ar, 4b - SiH_4-H_2-He; 7-10 - SiH_4-NH_3-N_2.

Generalized temperature dependence of the concentrations of hydrogen for methods M1, M3, M4 is presented in Figure 16 [157]. The general trend of the increase of the concentration of hydrogen with a fall of the temperature of deposition is obvious. The spread of the experimental points in the areas outlined by the dotted line for different methods of deposition (M1, M3, M4 are designated) is due to the differences in the compositions of the gas mixes and parameters of plasma (at a high frequency of excitation the concentration of hydrogen is up to one and a half times higher than at a low frequency). The presented data testify to the fact that because of a considerable content of hydrogen in SNTF, for the description of its real composition in a general form it is necessary to use $Si_xN_yH_z$ formula.

8.4. Basic Physical and Chemical Properties of SNTF

In literature various physical and chemical characteristics of SNTF are discussed. Most authors frequently analyze the methodically simple and accessible rate of dissolution of SNTF in the solutions based on hydrofluoric acid. According to different authors, the rates of dissolution demonstrate a trend to a decrease with the rise of the temperature of deposition of the films. Besides, an increase in the concentration of hydrogen results in higher wet etch rates [66]. However, the problems are because different authors have very different compositions of solvents, SNTF are obtained at various temperatures, etching is influenced strongly by the compositions of the films and conditions for etching, etc. Thus, a technologically convenient characteristic of SNTF can be considered only as of the secondary one, and this is of little use for the comparative analysis of the SNTF properties.

In this connection, the author for a comparative analysis made his choice in favor of the primary characteristic of SNTF – film density (γ), although this characteristic was investigated in a considerably smaller number of works. The value of density for S-SNTF is 2.9 – 3.1 g/cm^3, Figure 17, which is close to the known reference value for a volume material – 3.17 g/cm^3. The ranges of the density variation for the activated methods of deposition in Figure 17 show a trend to a decrease of the density of the low-temperature

SiNH-TF, which correlates with the trends for an increase of the concentration of hydrogen and enrichment with silicon (reference density of silicon - 2.33 g/cm^3). For the latter case, for example, in [67] a direct dependence is shown of the decrease of γ in the range of 3.0 – 2.0 during a decrease of Si/N in the range of 0.75 – 0.5. Naturally, the rates of dissolution of SiNH-TF demonstrate an inverse dependency on the density of the films [118].

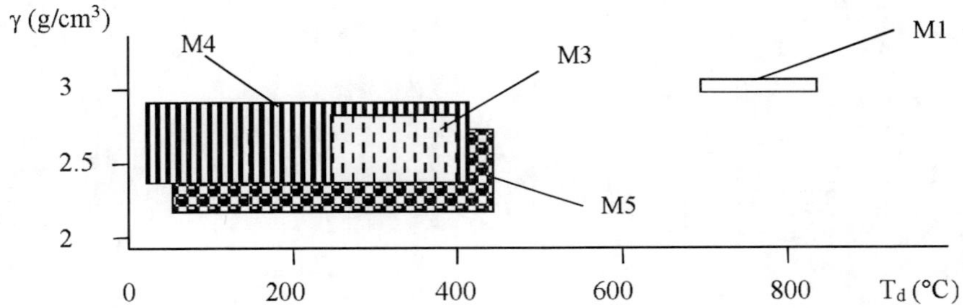

Figure 17. Exemplary ranges of SNTF density values obtained by methods M1, M3, M4, M5 with the corresponding approximate temperature intervals.

8.5. Basic Physical and Chemical Properties of SNTF

The temperature trends constructed according to the literature on the SNTF density for various methods of their preparation are provided in Figure 18; in the same place there are accepted average values of the density according to the data of generalizations [157]: method M1 – ~3.0 g/cm^3, method M3 – ~2.6 g/cm^3, method M4 – ~2.9 g/cm^3. It is possible to see an obvious trend for increasing the density up to the reference value of the volume of silicon nitride of 3.1 g/cm^3. In general, this trend is the reverse with the trend in Figure 16. An analysis of the literary data on the correlation of the influence of hydrogen on the density of the films is presented in Figure 19 and it shows an inverse dependence of the density on the concentration of hydrogen in SNTF. The reason for the variation of the density is the changes in the composition of SNTF. Let us note that in the analysis of the density of SiNH-TF it is necessary to take into account that their composition may be different by the Si/N ratio, while enrichment of the films with silicon also leads to a decrease of their density. It was established that the rate of dissolution of SNTF in the etchants based on hydrofluoric acid (V_{etch}) was proportional to [H]. Since it is significantly easier to control the rate of dissolution, there are many published experimental data, starting with the very first publications. In general, the numerous experimental data published in the work [15] correspond to the approximate ratio of $V_{etch} \sim 0.99^{0.23[H]}$ [99].

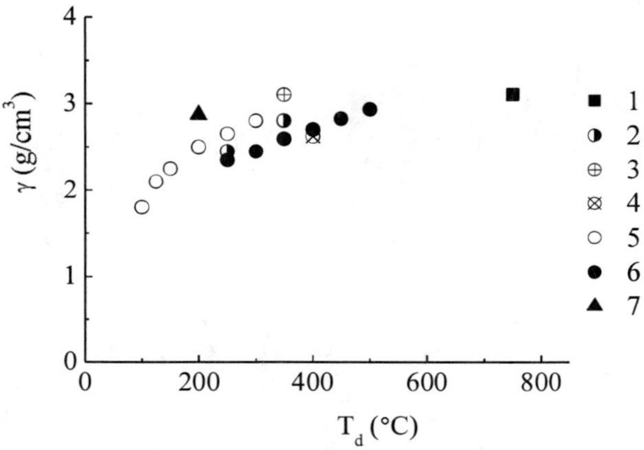

Figure 18. Temperature dependences of density in SNTF for various methods of their preparation [157]: 1 – method M1 (LPCVD); 2 – 6 – method M3 (PECVD); 7 – method M4 (HDP-CVD). The mixtures compositions according to the Method of 3: 3 – SiH_4-N_2, 4,5 – SiH_4-NH_3-N_2, 6 – SiH_4-NH_3.

The total of the analyzed data allowed the author to estimate the interrelation of the temperature of the beginning of the out-diffusion of hydrogen from a film (or changes of any physical properties) with the total concentration of hydrogen in SiNH-TF, Figure 20 [157]. Data are presented for a long furnace anneal (FA) and rapid thermal anneal (RTA) conditions. Data show the trend of the decrease of temperature at the beginning (T_b) of the out-diffusion of hydrogen during the increase of its concentration in SiNH-TF. Indeed, SiNH-TF from the initial [H] ~20 at % begins to emit hydrogen even at 400°C. Hydrogen emission ends at about 800°C. It would be logical to assume that such films are more subjected to structural transformations even at a very low temperature of annealing. This seems to explain the efficiency of the use of the version of the multilayered coverings of deposition with the intermediate annealing of the thin layers at the same temperatures. On the contrary, the films with a smaller content of hydrogen are more stable.

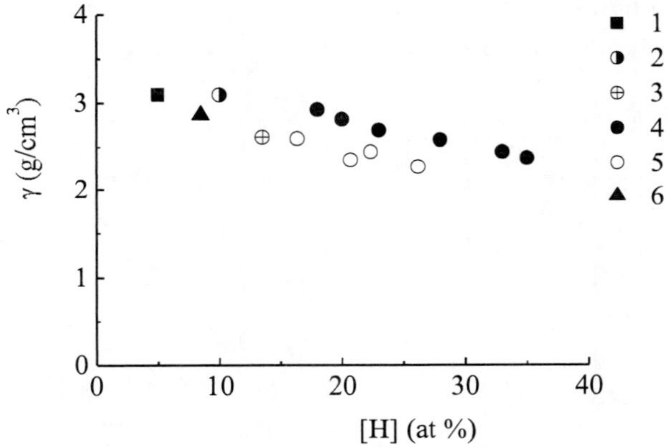

Figure 19. Correlation of SNTF density and hydrogen concentration in films [157]: 1 – method M1 (LPCVD); 2 – 5 – method M3 (PECVD); 6 – method M4 (HDP-CVD). The compositions of the mixtures according to the method of 3: 2 – SiH_4-N_2; 3,5 – SiH_4-NH_3-N_2, 4 – SiH_4-NH_3.

In conclusion of this subsection, we will point out that in many publications (see, summary [157]) attempts were made to explain the processes happening during the heating of SiNH-TF. Because of its availability and simplicity, the FTIR method was mainly used. In IR spectra the intensity of the peaks was studied (or the areas under these peaks) of Si–H bonds at ~2160 cm^{-1} and of N–H at ~3350 cm^{-1}. Then a comparison was carried out of the temperature dependences of variation of [Si–H] and [N–H] and, on their basis – of the total concentration of [H]. In the work [88] the partially deuterated SiNH-TF were also used due to application during the preparation of ND_3 films; respectively, besides Si–H and N–H, also peaks of Si–D, N–D were investigated at 2480 and 1585 cm^{-1}. Because of the multi-factor character of the methods for obtaining SiNH-TF, the modes of annealing and experimental analytical techniques, a detailed analysis of the question is beyond the limits of the present publication. Brief results of the research works can be presented as follows: at low temperatures of annealing the out-diffusion of the molecular hydrogen is limited by the concentration of [Si–H] bonds and is characterized by the formula of reaction ≡Si–H + =N–H → ≡Si–N= + H_2; at higher temperatures, this process goes faster, and already in absence of Si–H bonds the out-diffusion of the molecular hydrogen is caused by the formation in a film of the gaseous ammonia according to the formula: =N–H + 2(–H) → NH_3.

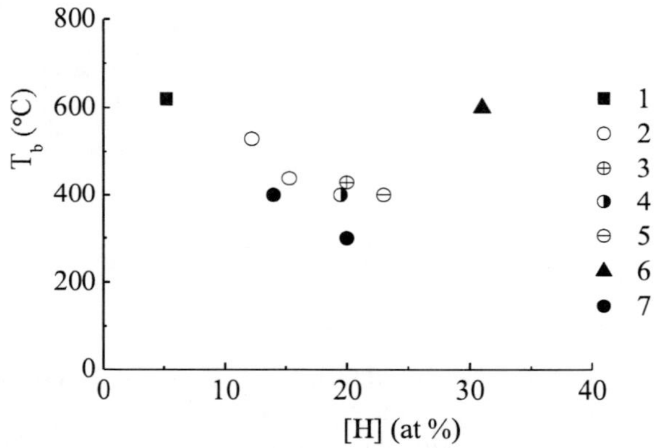

Figure 20. The relationship of the temperature of the onset of hydrogen evolution from SiNH-TF (T_b) with prolonged (FA) or rapid (RTA) thermal annealing with initial concentration of hydrogen in SiNH-TF [157]: 1,2,4,7 – FA; 3, 5, 6 – RTA. Sample 6, falling out of the trend, is a SiNH-TF enriched with silicon [158].

For MEMS technologies the most topical parameter of TF is the value of the mechanical stresses. By the results of generalizations [12, 38, 99, 152–154], for SNTF the sign and the value of the mechanical stresses depend on numerous factors, and in this connection, this question demands a separate discussion going beyond the framework of the present work. Reference ranges of the mechanical stresses (σ) for various methods of obtaining SNTF are presented in Figure 21. One can see that for S-SNTF the maximal values of the tensile, plus sign, mechanical stresses are typical. A version of a considerable decrease of the stress values for S-SNTF was demonstrated in [17, 19] through enrichment of a film with silicon and in the work [19] 1.5

times employing increasing the deposition temperature for SiH_4 – NH_3 mixture. For the former case, it is demonstrated by a block arrow in Figure 21 and the corresponding arrow in Figure 15(a) for the level of variation of Si/N correlation. On the contrary, for the activated processes in SiNH-TF, usually, a compressive (minus sign) type of stress is observed more likely. A comparison of the data in Figure 21 shows, that the greatest possibilities for variation of the sign and the values of the mechanical stresses are provided by method M3. For example, the results of the multifactorial experimental research work in [159] demonstrated, that the trend toward tensile stresses was observed at the rise of the deposition temperature and frequency of the plasma excitation and also at the rise of temperature of the thermal annealing, which followed the deposition.

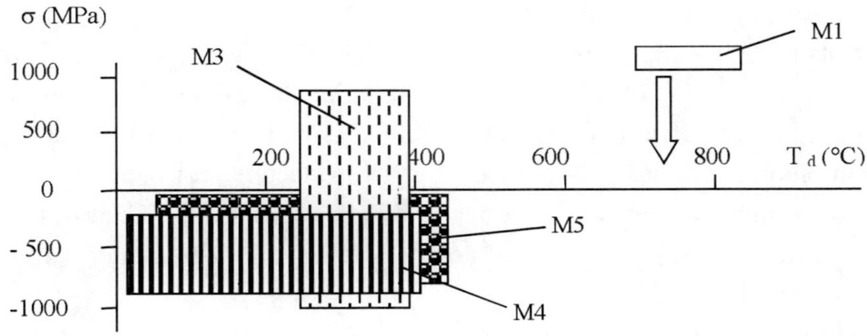

Figure 21. Exemplary ranges of mechanical stress values in SNTF, obtained by methods M1, M3, M4, M5 at the corresponding approximate temperature intervals.

8.6. Conclusion to Section 8

The state and directions for the development of six basic methods for obtaining SNTF for IC and MEMS applications were analyzed. Such methods are the deposition methods in the reactors of the flow and cyclic types during the thermal activation of the initial reagents, during the plasma activation, and also the method of the catalytic deposition with an application of a decomposition of the initial reagents on a hot wire. A detailed analysis of the technologies for obtaining, specific features and directions for the development of the concrete processes are presented in the corresponding sections. The original author reviews on the topic can be found elsewhere [12, 38, 39, 99, 152, 153, 154, 156, 157]. The analysis comprised the basic publications for several decades of the research works and included specific features of the equipment, comparison of the advantages and drawbacks of the applied chemical reagents, used parameters of the processes, etc. The generalized comparative data presented in part 7 of the review in the form of the temperature areas for the rates of deposition, composition, and basic properties of SNTF demonstrate the specific features of the analyzed methods of deposition. This makes possible not only a visual comparison of the basic characteristics of the deposition processes and TF parameters but also an estimation of the applicability of the basic methods of obtaining TF for solving concrete tasks in the microelectronic and MEMS technologies.

Data in Figure 22 summarizes revised SNTF CVD estimation necessary for the understanding of the studied deposition methods. This figure shows the dependences of thin film conformality (%) and the effectiveness of silicon compound use (%) concerning the effective constant of the deposition rate k_{eff}. It can be seen that the increase of k_{eff} is linked to the increase of the precursor's effectiveness. However, thin film conformality goes oppositely.

The author is confident that Figure 22 reflects the current understanding of SNTF film CVD features. It can be effectively used for the practical evaluation of the methods of CVD TF application. The choice of the SNTF CVD method will depend on the practical goal. For instance, highly activated CVD processes are suitable to deposit a relatively thick film on the flat substrate (without relief). In contrast, very slow ALD and LPCVD processes are necessary to obtain very thin films on structures with highly aggressive aspect ratios. This Figure indicated that at the current level on CVD/ALD understanding the deposition of conformal films with high deposition rates is impossible.

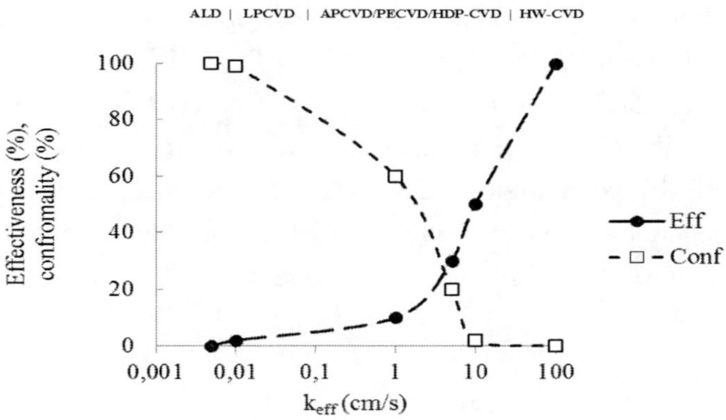

Figure 22. Estimation of dependences of TF conformality (%) and the effectiveness of silicon compound use (%) with respect to the effective constant of the deposition rate k_{eff}.

Conclusion

Table 9. Major SNTF deposition methods and film composition

CVD method	Major precursors	Nitrogen-contained reactants	Approximate intervals for deposition conditions and related basic film compositions			
			T_d (°C)	W_d (nm/min, CVD) (nm/cycle, ALD)	Si/N	[H] (at %)
LPCVD RTCVD	SiH_2Cl_2 $SiCl_4$ SiH_4	NH_3	700 – 850	1.5 – 4	0.75	< 5
TA-ALD	$SiCl_4$ SiH_2Cl_2 Si_2Cl_6 Si_3Cl_8	NH_3	350 – 600	< 0.2	0.75 – 1.0	-
PECVD	SiH_4	NH_3/N_2	250 – 400	20 – 200	0.8 – 1.2	20 – 40
HDP-CVD	SiH_4	N_2	25 – 400	15 – 90	0.6 – 1.0	5 – 15
PE-ALD	SiH_2Cl_2 Si_2Cl_6 SiH_4 $(SiH_3)_3N$ $(SiH_3)_4Si$ BTBAS	NH_3 (N_2, H_2)	50 – 450	< 0.15	0.55 – 1.2	10 – 15
HW-CVD	SiH_4	NH_3 (N_2, H_2)	230 – 500	20 – 250	0.7 – 1.5	10 – 15

Generalized data on CVD processes and silicon nitride film composition are listed in Table 9. Up to date, LPCVD, PECVD, and HDP-CVD are mature methods. They allow containing S-SNTF and SiNH-TF films with a wide range of film compositions and properties, which are effectively used in IC and MEMS technologies. ALD methods seem to be under intensive study with regards to their effectiveness (deposition rate, productivity, the effectiveness of reactant usage, etc.) and conformality.

REFERENCES

[1] Adams, AC. In *VLSI Technology*, 2nd edition, Sze SM. (Ed.), McGrow-Hill Int. Ed., 1988, pp 233-271.

[2] Vasilyev, VYu. *Thin Film Chemical Vapor Deposition in Integrated Circuit Technology: Equipment, Methodology and Thin Film Growth Experience* / Nova Science Publishers, Inc., 2013, pp 1-314.

[3] Vasilyev VYu. Nano- i mikrosistemnaja tehnika, 2014, no.9 (170), 37-44 (In Russian).

[4] Morosanu, CE. *Thin Solid Films*, 1980, vol.65, no.1, 171-208.

[5] Rzhanov, AV (Ed.) Silicon Nitride in Electronics, (Novosibirsk: Nauka, 1982).

[6] Stoffel, A.; Kovacs, F.; Kronast, W.; Muller, B. *J. Micromech. Microeng.*, 1996, vol.6, 1-13.

[7] Joshi, BC.; Eranna, G.; Runthala, DP.; et al. *Indian J. of Eng. & Mater. Sci.*, 2000, vol.7, no.4, 303-309.

[8] Xin, M.; Byun, Y.-C.; Kim, H.S.; et al. *Materials*, 2016, no.9, 1007.

[9] Kaloyeros, AE.; Jove, FA.; Goff, J.; et al. *ECS J. of Solid State Sci. and Techn.*, 2017, vol.6, no.10, P691-P714.

[10] Knolle, WR.; Osenbach, JW. *J. Appl. Phys.*, 1985, vol.58, no.3, 1248-1254.

[11] Emmanuel, NM.; Knorre, DG. *Treatise on Chemical Kinetics*, Moscow, 1984 (In Russian).

[12] Vasilyev, VYu. *Nano- i Mikrosistemnaya Tekhnika*, 2018, vol.20, no.5, 287-296 (In Russian).

[13] Gumpher, J.; Bather, WA.; Wedel, D. *IEEE Trans. Semicond. Manuf.*, 2003, vol.16, no.1, 26-35.

[14] Williams, K.; Brenniman, E.; Su, J.; et al. *Solid State Technol.*, 2007, vol.50, no.8, 41–44, 63.

[15] Bagatur'yants, AA.; Novoselov, KP.; Safonov, AA., et al. *Surface Sci.*, 2001, vol.486, 213-225.

[16] Makhviladze, TM.; Minushev, AK. *Proc. of SPIE*, 2010, vol.7521, 75211A1-75211A12.

[17] Gardeniers, JGE.; Tilmans, HAC.; Visser, CCG. *J. Vac. Sci. Technol. A.*, 1996, vol.14, no.5, 2879-2892.

[18] Zheng, B.; Zhou, C.; Wang, Q.; et al. *Advances in Materials Science and Engineering*, 2013, vol.2013, # 835942.

[19] Temple-Boyer, P.; Rossi, C.; Saint-Etienne, E.; et al. *J. Vac. Sci. Technol. A.,* 1998, vol.16, no.4, 2003-2007.

[20] Walker, KL.; Jardine, RE.; Ring, MA.; et al. *Int. Journal of Chemical Kinetics,* 1988, vol.30, 69-88.

[21] Walch, SP.; Dateo, CE. *J. Phys. Chem. A.,* 2001, vol.105, 2015-2022.

[22] Osenbach, JW.; Zell, JL.; Knolle, WR.; et al. *J. Appl. Phys.,* 1990, vol.67, no.11, 6830-6843.

[23] Rosler, RS.; Engle, GM. *Solid State Technol.*, 1981, vol.24, no.4, 172-177.Gsd

[24] Levy, RA.; Lin, X.; Grow, JM. *J. Mat. Res.,* 1996, vol.11, no.6, 1483-1488.

[25] Yeh, WC.; Ishihara, R.; Morishita, S.; et al. *Jpn. J. Appl. Phys.,* 1996, vol.35, part 1, no2B, 1509-1512.

[26] Lin, X.; Endisch, D.; Chen, X.; Kaloyeros, A. *Mat. Res. Soc. Symp. Proc.,* 1998, vol.495, 107-112.

[27] Jhansirani, K.; Dubey, RS.; More, MA.; et al. *Results in Physics*, 2016, vol.6, 1059-1063.

[28] US Patent Application 2006/0102076, 2006.

[29] US Patent Application 2002/0045362, 2002.

[30] Borisenko, VE.; Hesketh, PJ. *Rapid Thermal Processing of Semiconductors*, Springer Science + Business Media, LLC, New York, 1997, 253-300.

[31] Wang, CY.; Lim, EH.; Liu, H.; et al. *Mat. Res. Soc. Proc., 2001*, vol.664, A8.5.1–A.8.5.6.

[32] Ritala, M.; Leskelä, M. In *Handbook of Thin Film Material,* H. S. Nalwa, Editor, p. 103, Academic Press, San Diego (2001).

[33] Putkonen, M.; Niinistö, L. *Top Organomet. Chem*., 2005, vol.9, 125.

[34] Edited by Hwang, CS.; Yoo, CY. *Atomic Layer Deposition for Semiconductors*, Springer, 2014.

[35] Kääriäinen, T.; Cameron, D.; Käriäinen, ML, Sherman, A. *Atomic Layer Deposition. Principles, Characteristics, and Nanotechnology Applications*. 2nd Edition, 2013, Scrivener Publishing LLC.

[36] Ovanesyan, RA.; Filatova, EA., Elliot, SD.; et al. *J. Vac. Sci. Technol.* A, 2019, vol.37, #060904.

[37] Knoops, HCM.; Langereis, E.; Van de Sanden, MCM.; et al. *J. Electrochem. Soc*., 2010, vol.157, no.12, G241-G249.

[38] Vasilyev, VYu. *Nano- i Mikrosistemnaya Tekhnika*, 2018, vol.20, no.6, 329-339 (In Russian).

[39] Vasilyev, VYu. *Electronic Engineering. Series 3. Microelectronics*, 2020, iss.1(177), 31-41 (In Russian).

[40] Everstein, FC.; Put, BN. *J. Electrochem. Soc*., 1973, vol.120, no.1, 106-110.

[41] Klaus, JW.; Ott, AW.; Dillon, AC.; et al. *Surface Science*, 1998, vol.418, L14–L19.

[42] Ooba, K.; Nakashima, Y.; Nakajima, A.; et al. *Ext. Abstr. 1998 Int. Conf. Solid State Dev. Mater.*, 1998, Hiroshima, 22-23.

[43] Nakajima, A.; Yoshimoto, T.; Kidera, T.; et al. *Appl. Phys. Let.*, 2000, vol.77, no.18, 2855-2857.

[44] Nakajima, A.; Yoshimoto, T.; Kidera, T.; et al. *Appl. Phys. Let.*, 2001, vol.79, no.5, 665-667.

[45] Nakajima, A.; Khosru, QDM.; Yoshimoto, T.; et al. *Appl. Phys. Let.*, 2002, vol.80, no.7, 1252-11254.

[46] Nakajima, A.; Khosru, QDM.; Yoshimoto, T., et al. *J. Vac. Sci. Technol. B.*, 2002, vol.20, no.4, 1406-1409.

[47] Nakajima, A.; Khosru, QDM.; Yoshimoto, T.; et al. *Microelectronics Reliability*, 2002, vol.42, 1823-1835.

[48] Nakajima, A.; Khosru, QDM.; Yoshimoto, T.; et al. *Appl. Phys. Let.*, 2003, vol.83, no.2, 335-337.

[49] Zhu, S.; Nakajama, A. *Jap. J. Appl. Phys.*, 2007, vol.46, no.12, 7699-7701.

[50] Lee, WJ.; Lee, JH.;, Park, CO.; et al. *J. Kor. Phys. Soc.*, 2004, vol.45, no.5, 1352-1355.

[51] Lee, WJ.; Kim, UJ.; Han, CH.; et al. *J. Kor. Phys. Soc.*, 2005, vol.47, no.11, S598-S602.

[52] Goto, H.; Shibahara, K.; Yokoyama, S. *Appl. Phys. Let.*, 1996, vol.68, no.23, 3257-3259.

[53] Yokoyama, S.; Ikeda, N.; Kajikawa, K.; et al. *Appl. Surf. Sci.*, 1998, vol.130-132, 352-356.

[54] Morishita, S.; Sugahara, S.; Matsumura, M. *Appl. Surf. Sci.*, 1997, vol.112, 198-204.

[55] Park, K.; Yun, WD.; Choi, BJ.; et al. *Thin Solid Films*, 2009, vol.517, 3971-3978.

[56] Granneman, E.; Fischer, P.; Pierreux, D.; et al. *Surf. Coat. Technol.*, 2007, vol.201, 8899-8907.

[57] Riedel, S.; Sundqvist, J.; Gumprecht, T. *Thin Solid Films*, 2015, vol.577, 114-118.

[58] Cheng, HC. In *ULSI Technology.* Chang CY., Sze SM. (Eds.), McGrow-Hill Int. Ed, 1996, pp 205-269.

[59] Ting, CH. In *Handbook of Multilevel Metallization for Integrated Circuits*. Wilson SR., Tracy CJ., Freeman JL. (Eds.), Noyes Publications, Park Bridge, NJ, 1993, pp 202-273.

[60] Shepelev, SN.; Vasilyev, VYu.; Popov, VP. *Obzory po jelektronnoj tehnike.* Ser.2. Vyp.7. Moscow, CNII "Jelektronika," 1986, 1-53 (in Russian).

[61] Shepelev, SN.; Vasilyev, VYu., Popov, VP. *Obzory po jelektronnoj tehnike*. Ser.2. Vyp.7. Moscow, CNII "Jelektronika," 1988, 1-62 (in Russian).

[62] Xia, LQ.; Chang, M. *In Handbook of Semiconductor Manufacturing Technology, second edition*. Ed. by Doering, R. and Nishi, Y., 2008, CRC press, p.p.13-1 – 13-87.

[63] Cote, DR.; Nguyen, SV.; Stamper, AK.; et al. *IBM J. of Res. Develop.*, 1999, vol.43, no.1/2, 5-38.

[64] Hamedani, Y.; Macha, P.; Bunning TJ.; et al. In *Chemical Vapor Deposition - Recent Advances and Applications in Optical, Solar Cells and Solid State Devices.* Ed. by Neralla S., 2016, InTex, 247-280.

[65] Stein, HJ.; Wells, VA.; Hampy, RE. *J. Electrochem. Soc.*, 1979, vol.126, no.10, 1750-1754.

[66] Chow, R.; Lanford, WA.; Wang, KM.; et al. *J. Appl. Phys.*, 1982, vol.53, no.8, 5630-5633.

[67] Huang, H.; Winchester, KJ.; Suvorova, A.; et al. *Mat. Sci. and Eng. A.*, 2006, vol.435-436, 453-459.

[68] Yan, HY.; Ou, KS.; Chen, KS. *Strain*, 2008, vol.44, 259-266.

[69] Jonak-Auer, I.; Meisels, R.; Kechar, F. *Infrared Phys. and Technol.*, 1997, vol.38, 223-226.

[70] Kim, NH.; Seo, YJ.; Park, SW.; et al. *Microelectron. Eng.*, 2005, vol.81, 29-34.

[71] Applied Materials, Inc. Producer® Celera™ PECVD. http://www.appliedmaterials.com/products/producer-celera-pecvd.

[72] Cotler, TJ.; Chapple-Sokol, J. *J. Electrochem. Soc.*, 1993, vol.140, no.7, 2071-2075.

[73] Wang, CY.; Lim, EH.; Vassiliev, VY.; et al. *Proc. 17th Intern. VLSI Multilevel Interconnection Conf. (VMIC)*, Santa Clara, California (USA), June 2000, 411-413.

[74] Allaert, K.; Casler, AV.; Loos, H.; et al. *J. Electrochem. Soc.*, 1985, vol.132, no.7, 1763-1766.

[75] Flamm, DL.; Chang, CP.; Ibottson, DE.; et al. *Solid State Technol.*, 1987, vol.30, no.3, 43,44.

[76] Zhang, X.; Grigoropoulos, CP. *Rev. Sci. Instrum.*, 1995, vol.86, no.2, 1115-1120.

[77] Hansch, W.; Nakajima, A.; Yokoyama, S. *Appl. Phys. Lett.*, 1999, vol.75, no.11, 1535-1537.

[78] Vila, M.; Caceres, D.; Prieto, C. *J. Appl. Phys.*, 2003, vol.94, no.12, 7868-7873.

[79] Onishaev, DYu. *Ph.D. dissertation synopsis*, Moscow, 2007, 22 pages (In Russian).

[80] Sharma, V. *Ph.D. dissertation,* Arizona State University, USA, 2013, 163 pages.

[81] Komarovy, FF.; Vlasukova, LA.; Parkhomenko, IN.; et al. *Proc. Int. Conf. Nanomaterials: Application and Properties*, 2013, vol.2, no.1, # 01TEA12.

[82] Boehme, C.; Lucovsky, G. *J. Vac. Sci. Technol. A.*, 2001, vol.19, no.5, 2622-2628.

[83] Nallapati G. *LSU Historical Dissertations and Theses*, 1999, Louisiana State University, #6899.

[84] Batan, A.; Franquet, A.; Vereecken, J.; et al. *Surf. Interface Anal.*, 2007, vol.40, 754-757.

[85] Grigaitis, T.; Naujokaitis, A.; Tumėnas, S.; et al. *Lithuan. J. Phys.*, 2015, vol.55, no.1, 35-43.

[86] Vassiliev, VY.; Sudijono, JL.; Cuthbertson, A. *Solid State Technol.*, 2001, vol.44, no.3, 129-136.

[87] Sato, M.; Yoshinobu, A. *J. J. Appl. Phys.*, 1986, vol. 256, no.9, L764-L766.

[88] Kotecki, DE.; ChappleSokol, JD. *J. Appl. Phys.*, 1995, vol.77, no.3, 1284-1293.

[89] Moshkalyov, SA.; Diniz, JA.; Swart, JW.; et al. *J. Vac. Sci. Technol. B.*, 1997, vol.15, no.6, 2682-2687.

[90] Xie, L.; Deng, J.; Shepard, S., et al. *Proc. NSTI-Nanotech Conf.*, 2005, vol.2, 328-331.

[91] Farber, DG.; Bae, S.; Okandan, M.; et al. *J. Electrochem. Soc.*, 1999, vol.146, no.6, 2254-2257.

[92] Lapeyrade, M.; Besland,,MP.; Meva'a, C.; et al. *J. Vac. Sci. Technol. A.*, 1999, vol.17, no.2, 433-444.

[93] Besland, MP.; Lapeyrade, M.; Delmotte, F.; et al. *J. Vac. Sci. Technol. A.*, 2004, vol.22, no.5, 1963-1970.

[94] Han, SS.; Jun, BH.; No, K.; et al. *J. Electrochem. Soc.*, 1998, vol.145, no.2, 652-658.

[95] Yota, J.; Hander, J.; Saleh, AA. *J. Vac. Sci. Techn. A.*, 2000, vol.18, no.2. 372-376.

[96] Lee, JW.; Mackenzie, KD.; Johnson, D.; et al. *J. Electrochem. Soc.*, 2000, vol.147, no.4, 1481-1486.

[97] Kim, H.; Lee, Y.; Ra, Y.; et al. *ECS Trans.*, 2007, vol.6, no.3, 531-547.

[98] Parm, IO.; Kim, K.; Lim, DG.; et al. *Solar Energy Materials & Solar Cells*, 2002, vol.74, 97-105.

[99] Vasilyev, VYu. *Nano- i Mikrosistemnaya Tekhnika*, 2018, vol.20, no.9, 542-554 (In Russian).

[100] Ma, TP. *IEEE Trans. Electron. Dev.*, 1998, vol.45, no.3, 680-690.

[101] Zambom, LS.; Mansano, RD.; Furlan, R.; et al. *Thin Solid Films*, 1999, vol.343-344, 299-301.

[102] Yasui, K.; Nasu, M.; Kaneda, S. *J. J. Appl. Phys.*, 1990, vol. 29, no.12, 2822-2823.

[103] US Patent Application 2014/0187045, 2013.

[104] George, SM. *Chem. Rev.*, 2010, vol.110, no.1, 111-131.

[105] Profijt, HB.; Potts, SE.; Van de Sanden, MCM.; et al. *J. Vac. Sci. Technol. A.*, 2011, vol.29, no.5, 050801-050826.

[106] Kessels, E.; Profijt, H.; Potts, S.; et al. In *Atomic Layer Deposition of Nanostructures Materials* (Pinna, N. and Knez, M., Ed.). 2012, Willey-VCH. Verlag, 131-158.

[107] Potts, SE.; Kessels, WMM. *Coord. Chem. Rev.*, 2013, vol. 257, 3254-3270.

[108] Heil, SBS.; Hemmen, J.L.; Hodson, CJ.; et al. *J. Vac. Sci. Technol. A*, 2007, vol.25, no.5, 1357-1366.

[109] King, SW. *ECS Trans.,* 2010, vol.32, no.2, 365-373.

[110] King, SW. *J. Vac. Sci. Technol. A,* 2011, vol.29, no.4, 041501/1-9.

[111] Week, S.; Nowling, G.; Fuchigami, N.; et al. J. Vac. Sci. Technol. A, 2016, vol.34, no.1, 01A140/1–6.

[112] Triyoso, DH.; Jaschke, V.; Shu, J.; et al. *Proc. 2012 Int. Conf. on ALD*, # 12851120.

[113] Triyoso, DH.; Hempel, K.; Ohsiek, S.; et al. *ECS J. of Sol. State. Sci. Technol.*, 2013, vol.2, no.11, N222-N227.

[114] Jang, W.; Jeon, H.; Kang, C.; et al. *Phys. Status Solidi A,* 2014, vol.211, no.9, 2166-2171.

[115] Jang, W.; Jeon, H.; Song, H.; et al. *Phys. Status Solidi A,* 2015, vol.212, no.12, 2785-2790.

[116] Park, JM.; Jang, SJ.; Yusup, L.; et al. *ACS Appl. Mater. Interfaces,* 2016, vol.8, no.32, 20865-20871.

[117] Yokoyama, S.; Goto, H.; Miyanoto, T.; et al. *Appl. Surf. Sci.*, 1997, vol.112, 75-78.

[118] Provine, J.; Schindler, P.; Kim, Y.; et al. *AIP Advances*, 2016, vol.6, no.6, 065012-065021.

[119] Ando, T.; Ohta, Y.; Ashihara, H.; et al. *ECS Trans.*, 2007, vol.11, no.7, 55-60.

[120] Nagata, K.; Nagasaka, M.; Yamaguchi, T.; et al. *ECS Trans.*, 2013, vol.53, no.3, 51-56.

[121] Koehler, F.; Triyoso, DH.; Hussain, I.; et al. *IOP Conf. Series: Mat. Sci. and Eng.*, 2012, vol.41, # 012006.

[122] Koehler, F.; Triyoso, DH.; Hussain, I.; et al. *Phys. Status Solidi C*, 2014, vol.11, no.1, 73-76.

[123] Ovanesyan, RA.; Hausmann, DM.; Agarwal, S. *ACS Appl. Mater. Interfaces*, 2015, vol.7, no.202, 10806-10813.

[124] Yusup, L.; Park, JM.; Noh, YH.; et al. *RSC Adv.*, 2016, vol.6, 68515-68524.

[125] Suntola, T. *Thin Solid Films*, 1992, vol.216, 84-89.

[126] Knoops, HCM.; Braeken EMJ,; de Peuter K.; et al. *ACS Appl. Mater. Interfaces*, 2015, vol.7, no.35, 19857-19862.

[127] Knoops, HCM.; de Peuter K.; Kessels, WMM. *Appl. Phys. Lett.*, 2015, vol.107, no.1, 014102–014106.

[128] Ande, CK., Knoops, HCM.; de Peuter K.; et al. *J. Phys. Chem. Lett.*, 2015, vol.6, no.18, 3610-3614.

[129] Bosch, RHEC.; Cornelissen, LE.; Knoops, HCM.; et al. *Chem. Mater.*, 2016, vol.28, no.16, 5864-5871.

[130] Andringa, AM.; Perrotta, A.; de Peuter P.; et al. *ACS Appl. Mater. Interfaces*, 2015, vol.7, no.40, 22525-22532.

[131] Gumpher, J.; Bather, W.; Mehta, N.; et al. *J. Electrochem. Soc.,* 2004, vol.151, no.5, G353-G359.
[132] US Patent Application 2017/0053792, 2016.
[133] Matsumura, H. *J. J. Appl. Phys.*, 1989, vol.28, no.10, 2157-2161.
[134] Schropp, REI. *Thin Solid Films*, 2015, vol.595, Part B (B), 272-283.
[135] Keum, KS.; Song, TH.; Hong, WK. *J. of Information Display*, 2014, vol.15, no.1, 25-30.
[136] Dupuie, JL.; Gulari, E.; Terry, F. *J. Electrochem. Soc.*, 1992, vol.139, no.4, 1151-1159.
[137] Deshpande, SV.; Dupuie, JL.; Gulari, E. *Appl. Phys. Lett.*, 1992, vol.61, no.12, 1420-1422.
[138] Deshpande, SV.; Gulari, E. *J. Appl. Phys.*, 1995, vol.77, no.12, 6534-6541.
[139] Deshpande, SV.; Dupuie, JL.; Gulari, E. *Appl. Mater. For Optics and Electronics*, 1996, vol.6, 135-146.
[140] Okada, S.; Matsumura, H. *Proc. Mat. Res. Soc. Symp.*, 1997, vol.446, 109-114.
[141] Okada, S.; Matsumura, H. *J. J. Appl. Phys.*, 1997, vol.36, Part 1, no.11, 7035-7040.
[142] Sato, H.; Izumi, A.; Matsumura, H. *Appl. Phys. Lett.*, 2000, vol.77, no.17, 2752-2754.
[143] Stannowski, B.; Rath, JK.; Schropp, REI. *J. Appl. Phys.*, 2003, vol.93, no.5, 2618-2625.
[144] Goldbach, HD.; Werf, CHM.; Löffer, J.; et al. *Conf. Record of the Thirty-first IEEE Photovoltaic Specialists Conference*, 2005, 1249-1252.
[145] Atwater, HA. Si Passivation and Chemical Vapor Deposition of Silicon Nitride. *Final Technical Report*, California Institute of Technology, 2007, 33 pages.
[146] Noh, SM.; Hong, WS. *J. Korean Cer. Soc.*, 2015, vol.52, no.3, 209-214.
[147] Schropp, REI.; Werf, CHM.; Verlaan, V.; et al. *Thin Solid Films*, 2009, vol.517, 3039-3042.
[148] Schropp, REI. *ECS Trans.*, 2009, vol.25, no.8, 3-14.

[149] Duan, HL.; Zaharias, GA.; Bent, SF. *Appl. Phys. Lett.*, 2001, vol.78, no.12, 1784-1786.
[150] Holt, JK. *Ph.D. dissertation*, California Institute of Technology, USA, 2003. 188 pages.
[151] Eustergerling, BD.; Shi, Y. *ARKIVOC*, 2009, Part (v), 75-89.
[152] Vasilyev, VYu. *Nano- i Mikrosistemnaya Tekhnika*, 2019, vol.21, no.1, 3-13 (In Russian).
[153] Vasilyev, VYu. *Nano- i Mikrosistemnaya Tekhnika*, 2019, vol.20, no.10, 585-595 (In Russian).
[154] Vasilyev, VYu. *Nano- i Mikrosistemnaya Tekhnika*, 2019, vol.20, no.11, 659-675 (In Russian).
[155] Semmache, B.; Lemiti, M.; Chaneliere, Ch.; et al. *Thin Solid Films*, 1997, vol.296, no.1-2, 32-36.
[156] Vasilyev, VYu. *Nano- i Mikrosistemnaya Tekhnika*, 2019, vol.21, no.3, 131-142 (In Russian).
[157] Vasilyev, VYu. *Nano- i Mikrosistemnaya Tekhnika*, 2019, vol.21, no.6, 352-367 (In Russian).
[158] He, SS.; Shannon, VL. *Proc. 4th Int. Conf. on Solid-State and IC Technol.*, 1995, 269-271.
[159] Claasen, WAP. Plasma Chem. Plasma Proc., 1987, voll 7, no.1, 109-124.

BIOGRAPHICAL SKETCHES

Vladislav Yu. Vasilyev, PhD

Affiliation: Novosibirsk State technical University, part-time professor at Semiconductors and Microelectronics Department, Novosibirsk, Russia.

Education: MS in chemistry from Novosibirsk State University in 1976, PhD in physical chemistry, and DSc in solid state chemistry from Russian Academy of Sciences in 1990 and 2002, respectively.

Business Address: Novosibirsk State Technical University, 20 Prospekt K. Marksa, Novosibirsk, 630073, Russia.

Research and Professional Experience: In total, Dr. Vasilyev has over 40 years of experience in microelectronics technology. He has more than 200 published scientific and technical papers and presentations at the National and International Conferences. He was granted with 15 USSR/Russian and 7 US patents. He is an active member of The Electrochemical Society.

Professional Appointments:

Last 20 years:

- 1996-2000: R&D department of Chartered Semiconductor Manufacturing, Ltd., principal engineer responsible for thin film development for various CMOS logic applications.
- 2001-2005: leading research scientist at the Institute of Semiconductor Physics, Siberian Branch of Russian Academy of Sciences.
- 2006-2008: Invited Professor for research under Korean Governmental Brain Pool Program at the Korea Polytechnic University, Department of Nano-Optical engineering.

2006 - present:

Deputy General Director of a sub-micron IC design center SibIS LLC, Novosibirsk, and since 2005: a part-time Professor at the Novosibirsk State Technical University, Semiconductors and Microelectronics Department.

Publications from the Last 3 Years:

- 2019: Vasilyev V. Yu. "Estimating the Interrelation between the Rate of Atomic Layer Deposition of Thin Platinum-Group Metal Films and the Molecular Mass of Reactant Precursors," *Russian Microelectronics* vol. 48, No 4, p.p.208-219 (DOI: 10.1134/S1063739719040103).
- 2020: Vasilyev V. Yu., Kittilsland G. "Review – Mechanical Stress in Chemically Vapor Phase Deposited Boron- and Phosphorus-Contained

Silicate Glass Thin Films: A Review," *ECS Journal of Solid State Science and Technology,* 2021, vol.9, # 043003 (DOI: 10.1149/2162-8777/ab8d93).

- 2021: Vasilyev V. Yu. "Review – Atomic Layer Deposition of Silicon Dioxide Thin Films," *ECS Journal of Solid State Science and Technology,* 2021, vol. 10, #053004, (DOI: 10.1149/2162-8777/abffab).
- 2019-2021: some papers on thin film topics in Russian scientific and technical journals:
    - DOI: 10.17587/nmst.21.40-51,
    - DOI: 10.17587/nmst.21.231-246,
    - DOI: 10.17587/nmst.21.298-309;
    - DOI: 10.7868/S2410993220010030,
    - DOI: 10.7868/S2410993220010042,
    - DOI: 10.7868/S2410993220030033,
    - DOI: 10.7868/S2410993220030045

In: Chemical Vapor Deposition (CVD) ISBN: 978-1-53619-949-9
Editor: Levi Karlsson

Chapter 2

Rapid-Heating Criterion in CVD Diamond Growth Technology and Possibility of Synthesizing Large Diamonds without High Pressure and Prepared Substrate

Nickolay I. Alekseyev*, Anton P. Broyko, Vladislav S. Khadutin, Ivan K. Khmelnitsky and Ivan V. Oreshko

Department of Micro and Nanoelectronics,
Saint Petersburg Electrotechnical University "LETI,"
St. Petersburg, Russia

Abstract

The existing technologies of single-crystal diamond synthesis for electronics are expensive and have significant limitations. For example,

* Corresponding Author's Email: nialekseyev@yandex.ru.

artificial diamonds obtained at high pressure and high temperature (HPHT), require high energy consumption, have sectorial structure and metal catalyst inclusions. CVD-synthesis of such diamonds requires single crystal substrate and does not provide overgrowth with maintaining high quality of the diamond. At the same time, there is a factor which is known in the known diamond synthesis technologies, but is not explicitly taken into account in theoretical models. It is the temperature rise rate in space or on time. In the 2000s this factor (hereafter fast-heating criterion) was proposed as an empirical criterion able to propose new methods of the diamond synthesis, free of both high pressure and a previously prepared substrate.

In this chapter we try to trace the mechanism underlying this criterion, in particular case of the CVD diamond growth technologies, enhanced by activation of the gas flow (MWCVD, HFCVD), and to assess the prospects of the old idea at a new level.

Analysis of the fast heating shows, that the density of the nucleation source which is a vapour of small carbon species formed from methane-carbon-oxygen mixture, follows the temperature rise with a much smaller delay than the density of the same species as the saturated vapour components above the cluster surface.

Therefore, the synthesis of the carbon clusters of both graphite and diamond types occurs under conditions of supersaturation of the small hydrocarbon vapor, initially enormous but rapidly decreasing then. The supersaturation is much higher in the diamond channel than in the graphene (or graphite) one. In addition, in the presence of an oxidizing agent, oxidation of the diamond nuclei occurs much lower than that of the graphite nuclei. These results are associated with the fact that both oxidation and desorption of the carbon species from the surface of the diamond clusters are accompanied by breaking larger number of carbon bonds; however, the difference in the oxidation factor is much stronger than the difference in the desorption factor.

In the above consideration of the MWCVD process, the substrate is involved only as a factor which determine (in combination with the plasma) the temperature dynamics in space. The diamond nuclei arising in the gas phase are deposited on the substrate as already formed clusters. However, the fast-heating criterion plays an important role in the CVD diamond synthesis technology anyway. Therefore, considering the ideology of CVD synthesis of high-quality massive diamonds to be fundamentally exhausted, this criterion can be applied to other possibilities of obtaining diamonds. The idea of non-stationary temperature dynamics and ability to use it to control the channels of gas-phase condensation can be realised under the synthesis not on a substrate, but in the bulk of a medium, which excludes free diamond growth.

Such a medium may be a carbonate-silicate melt, in which the carbon cluster tends to develop as the diamond cluster rather than the graphite one.

In this case experimental conditions are possible under which the fast-heating criterion is valid in the melt layer through which the gas mixture is passed.

The most significant of these conditions such as the necessary nucleation temperature of the diamond clusters and the temperature rise rate from the initial melting temperature to this nucleation temperature, are achievable under nonstationary HF induction heating at moderate parameters.

The values of these parameters practically do not differ from that of the MWCVD process, with the same simplifications that were used to describe the MWCVD. The nucleation temperature is about 1800-1900 K, but this value is based on a number of energy parameters obtained from quantum chemistry calculations and is estimated with an accuracy of ±400 K. So, the window of both the cluster nucleation and survival of the clusters is rather narrow and should be determined experimentally.

Keywords: CVD synthesis of diamonds, HPHT (high pressure - high temperature) synthesis of diamonds, nucleation of new phase

INTRODUCTION

Diamond is a material exceptional in its electronic and mechanical characteristics, and a number of unique possibilities in the extreme-, power, microwave-, acoustic electronics are associated with it [1-4]. Although for a number of applications polycrystalline diamond obtained by CVD (chemical vapour deposition) technology with diamond nuclei on available substrates [2] is sufficient, single crystal diamond (SCD) is much more promising.

But large natural SCD diamonds are very expensive as well as artificial SCD-diamonds obtained at high pressure and temperature (HPHT method). In addition, they always have sectorial structure, inclusions of a metal catalyst, and other disadvantages.

The other possibility of synthesizing single-crystal diamonds, different from HPHT, is CVD growth on the basis of already existing SCD substrates. These substrates can be natural diamonds or HPHT-diamonds. Also, an inhomogeneous substrate variant such as yttrium substrate is known, but yttrium is almost as expensive as the diamond itself.

Besides that, the CVD method does not provide a qualitative growth of the SCD to a large size due to a number of factors. In particular, the growth of various faces into an unconstrained space adjacent to the substrate is highly dependent on the impurities, occurs at different rates in different directions, and ultimately always leads to the appearance of a polycrystalline "coat."

Truly, everything above applies to massive SCD. The problem of creating a monocrystalline diamond film seems to be closer to its solution. In 2018 the SCD film was synthesized on the basis of a monocrystalline film of a cubic silicon carbide polytype (3C-SiC) processed with gas-phase CF_4 [5].

The method proposed is similar to the synthesis of the very original SiC film from a silicon substrate treated with gas-phase CO [6-7]. A feature of the solid-phase substitution method [5-7] is that it is carried out without a prepared diamond-like crystal matrix and does not appeal to high pressure. This feature makes the technology [5] similar to the feature of many natural diamond deposits, which also do not bear traces of high pressure and, at the same time, apparently do not have a ready-made crystalline matrix [8].

Therefore, for synthesizing massive SCD it is natural to search a way, which could partially borrow the features of such natural mechanism. But determining the conditions for this synthesis is a difficult problem. The HPHT and traditional CVD techniques enhanced by microwave plasma (MWCVD) or by hot filament (HFCVD) confine the possibilities of the SCD synthesis, if one remains within the framework of the standard concept of the diamond growth [9-11].

In this concept graphite and diamond are two allotropic modifications of pure carbon. The graphite is perceived as a form more stable under normal conditions, stable diamond falls into high-pressure region and is considered metastable at low pressures. The boundary between the two modifications on the pT plane (Bundy's diagram [9]) is constructed theoretically on the basis of Henry's law, which implies the same chemistry of the oxidation reactions of graphite and diamond as of pure carbon forms. It follows from this diagram that the diamond is nucleated either in a high-pressure region (like natural diamond in the most common geological models), or on a finished diamond-like matrix (at least SCD), as in the CVD technology.

From our view point, thus concept is rather traditional than justified. There is an alternative concept [12, 13], in which the diamond is the only form of pure carbon. It is absolutely stable under normal conditions and does not transform into graphite without participation of a catalyst, of hydrogen or water vapor. The graphite in this concept is, in principle, a hydrocarbon and its size is constrained. Bundy's theoretical phase diagram, which includes various solid-state modifications of pure carbon, becomes incorrect in this concept, since chemically different substances are compared.

The concept of the non-necessity of high pressure and of purely chemical nature of the graphite-to-diamond transformation was also discussed in the review [8], where a good deal of phenomenological information on the diamond- and graphite polycondensation of hydrocarbons in a heterogeneous catalytic reactor is presented. Kinetic conditions can be created in this reactor, and they that direct the process (which, e.g., proceeds by a formal equation $2CO + H_2 \rightarrow C + H_2O$) along the diamond synthesizing path, not the graphite one.

Unfortunately, there are currently no exact conditions for polycondensation just along the diamond path. It is also difficult to reproduce a hypothetical Portnov's alternative mechanism of the diamond synthesis in nature [14], where optimal synthesis conditions arise under sharp change in the redox state of the gas fluid with pressure drop.

We think that the kinetic criterion proposed in [12-13] and reduced to the fact that the appearance of diamond nuclei (and not graphite ones) is associated with fast temperature rise in space or in time, seems to be more promising. Herewith, the temperature can be lower than necessary for the formation of graphite (the fast-heating criterion). The criterion is traced both in the HPHT setup, where the temperature increases at a rate of about 1000 K/ s, and in effective versions of the CVD technique (MWCVD and HFCVD), where the gas temperature also increases sharply when approaching the substrate, and then decreases [15-16].

We set out to analyse the meaning of this empirical criterion in terms of the new phase nucleation theory, at least for the CVD synthesis conditions.

The next step could be the concept of experimental conditions for the synthesis of large SCDs that would not require extreme parameters and, at the same time, would not be related to CVD.

The proposed consideration does not cancel the traditional theory for the formation of diamond films on the diamond- or non-diamond substrates in CVD processes, but complements it. Indeed, while the CVD growth theory on the diamond substrate is generally clear, the diamond nucleation on substrates incommensurate with diamond does not seem to be the only mechanism. Anyway, the diamond synthesis of nuclei in the gas phase before their deposition on the substrate can be considered as a parallel process.

1. General Physical Picture

To interpret the fast-heating criterion, the following physical mechanism is proposed and then justified. The graphene- and graphite nuclei (clusters) have higher binding energy per a carbon atom than the diamond clusters, i.e., they are more advantageous (Figure 1, the existence of the intersection point of the curves 1, 2 for the diamonds and the graphite, respectively, in a large region of carbon atoms number *g* is problematic). At the same time, the very concept of infinite graphite, which is included in the traditional model and is embedded in quantum-chemical programs to allow carbon clusters of any size be compared, is controversial, from our viewpoint.

Therefore, using these programs and not sharing the traditional model, we will assume that the diamond clusters, that have arisen when considering the nucleation kinetics, are the final result and do not require transformation from a "metastable" form of carbon to some other. In this case, we will simply say that the nucleation occurs in the diamond channel and not in the graphite channel.

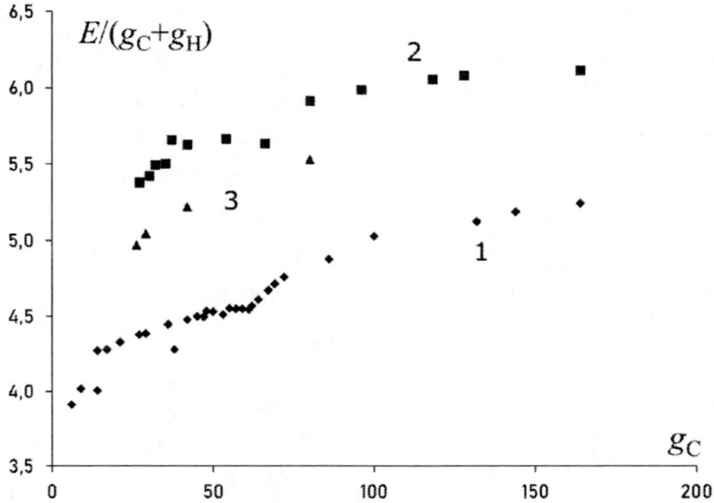

Figure 1. Binding energy (eV) per a carbon atom depending on a carbon atom number in the cluster for: 1 - diamond cluster; 2 - graphene cluster; 3 - bigraphene clusters. The curve 3 includes three points only. To determine the parameters of clusters, the package HyperChem [19] was used.

When using the gas mixture of methane plus hydrogen, the nucleation source is a vapour of small hydrocarbon species CH_3, $C_2H_{4\text{-}5}$, as well as of CH_2, CH, C. The supersaturation is determined by the ratio of the source vapour density to the vapour density of the same components desorbed from the surface of the carbon cluster. With a slow increase in temperature, this supersaturation is expressed through the difference in the binding energy during the synthesis of the cluster nucleation source and during the desorption from the cluster surface.

For the graphite and diamond, the formal external sequence of the desorption reactions and the components involved in it are almost identical. This is true both when oxygen is involved in such desorption and in its absence. However, in the diamond clusters, there is an additional act of breaking the C–C bond which takes place in the most of the desorption reaction sequence. So, the reaction sequences of the desorption are longer, as well as the bottleneck of such sequences.

Therefore, at a slowly varying temperature, the supersaturation in the diamond channel is characterized by a smaller positive or even negative exponential indicator, and the nucleation explosion occurs in the graphite channel. Then the supersaturation subsides and is never reached again.

A different situation is realized with a rapid increase in T. The increase in the saturated vapour density over the surface of the carbon clusters occurs much slower. Herewith the larger delay is observed just in the diamond channel, which is characterized by the longer reaction sequence and by the higher activation energy at the bottleneck of these sequences. So, greater supersaturation is formed in the diamond channel, and the contrast between two supersaturations increases with both temperature rise rate and oxygen density.

Clarification of this picture requires preliminary resolution of a number of auxiliary problems. These are outlined below. Note that when studying such a rough criterion as the fast-heating criterion, we tried to discard as much as possible the specific factors associated with plasma in the MWCVD (ions, charged radical ions, several different temperatures, etc.). They are essential, but in the method that is ultimately proposed and uses the fast-heating criterion, they can hardly be used.

2. Elementary Processes of Gas-Phase Clusters of Diamond and Graphite and Their Description Technique

2.1. Number of Surface Atoms in Diamond- and Graphite Clusters and Surface Energy of These Clusters

The simplest estimate of the exponent indicator in Zeldovich's formula [17-18], which describes the flow of emerging clusters through the bottleneck in the size space, has the form $-f=-\Delta G_c/\mathrm{T}=-(16/3)\left(\pi\sigma^3/\varsigma^2c^2\mathrm{T}^3\right)$ (hereinafter, the lower small sign "c" corresponds to the critical parameters). Here ς is the logarithm of supersaturation S (supersaturation), defined as the ratio $S=N_1/N_1^{(des)}$ of the gas-phase monomer density (the source of the diamond cluster growth) to the saturated vapour density of the same species desorbed from the cluster surface, c is the condensed phase density of atoms, σ is the surface energy per unit area. With standard CVD synthesis temperature of ~1000 K, σ ~ 11000 Erg/cm^2, the indicator value $|f|$ can be of the order of unity at giant supersaturations only, when the critical number of atoms in the cluster becomes <1 and there is no talk about the diamond or graphite at all.

However, this estimate is based on the assumption that the surface energy formed by the number g_0 of dangling valences of the surface carbon atoms is proportional to $g^{2/3}$ (where g is the total number of atoms in the cluster). This is not true if g is small, e.g., in the critical size cluster. So, the nucleation barrier is much lower and must be calculated more accurately.

We have built a number of gas-phase carbon clusters evolving from the smallest sizes. The evolution meant increase in the flat hexagonal cell number in the graphite channel and the path "armchair cyclohexane-to-adamantane-to-unit diamond cell" etc., in the diamond channel. Thus, we considered only the limiting variants of the C–H homology series [8], including only pure forms sp^3 and sp^2. To optimize configuration of the clusters and determine their thermodynamics, we used the simplest applied quantum-chemical package HyperChem [19].

Since real nuclei of both graphite and diamond under the CVD conditions are formed in a hydrogen-containing medium, the surface atoms of the simulated clusters were "loaded" with hydrogen atoms in the both cases, partially compensating for the surface energy of the clusters.

The construction of the diamond clusters with increasing number of atoms showed that the "bulky" (or internal) carbon atoms C, i.e., the atoms bonded with four neighbouring C atoms appear starting with $g \sim 20$. Then, focusing on the asymptotic $g_0 \to (36\pi)^{1/3} g^{2/3}$, it can be written:

$$g_0^{(\text{diam})} = \frac{g}{1+\left(g/36\pi\right)^{1/3}} \tag{1}$$

so that for $g \le 20\ g_0^{(\text{diam})} \approx g$. The factor $(36\pi)^{1/3}$ equals the density of the surface C-atoms with that of C-atoms in the "face" subsurface of the diamond surface (111). For the graphite, the shape of the $g_0(g)$ dependence is more complicated. For a small *g* (i.e., in the graphene), the number of atoms at the edge of the graphene plane can be written by analogy with (1):

$$g_0^{(\text{graph})} = \frac{g}{1+(g/4\pi)^{1/2}} \tag{2}$$

With increase in *g*, the graphene topology is replaced by the graphite one. Indeed, there is a certain cluster radius r_0, over which both cluster expansion (increase $r_0 \to r_0'$ with a doubling of the number of atoms) and transition to bigraphene are possible. In the first case, the binding energy change of the cluster is

$\left(R(r_0')-R(r_0)\right)E_{\text{rad}} - E_{\text{CC}}\left(\pi\left(r_0'^2-r_0^2\right)\right)/\,\text{a}_{\text{CC}}^2$, in the second

$R(r_0)E_{\text{rad}} - E_{\text{CC}}\left(\pi\left(r_0'^2-r_0^2\right)\right)/\,\text{a}_{\text{CC}}^2 - \left(\pi r_0^2/\,\text{a}_{\text{CC}}^2\right)E_{\|}$,

In this formula $R(r_0) \approx 2\pi\left(r_0' - r_0\right)/\,\text{a}_{\text{CC}}$ is the number of "radicals" at the edge of the cluster (by the radicals we mean carbon atoms having only two carbon neighbors and bonded with a hydrogen atom), $\text{a}_{\text{CC}} = 1.42$ Å is the C–C bond length in the graphene, the radical energy E_{rad} is the energy difference between the chemical C–C bond in graphene (~5.5 eV) and C–H bond (~4.05 eV), $E_{\|} = 0.04$ eV, interaction energy of a graphite monolayer with a parallel layer (determined from the HyperChem) per a carbon atom. If $r_0' \approx r_0\sqrt{2}$ is accepted for an estimation and the entropy contribution is not taken into account, the maximum graphene layer size is

$$\frac{r_0}{\text{a}_{\text{CC}}} \approx \frac{E_{\text{rad}}}{E_{\|}}\frac{1}{2\left(2-\sqrt{2}\right)} \sim \frac{E_{\text{rad}}}{E_{\|}} \tag{3}$$

Then the value obtained $(E_{rad}/E_{\parallel})$~40 corresponds to the number of one-layer atoms ~ 8000. The formula (3) can be easily generalized to N-layered graphite:

$$\frac{r_0}{\mathrm{a}_{CC}} \approx \frac{E_{rad}}{E_{\parallel}} 2\left(N+1-\sqrt{N(N+1)}\right) \approx \frac{E_{rad}}{E_{\parallel}}.$$

Thus, the growing graphite tends to form a stack of limited size layers to increase the number of layers, rather than their diameter.

The number of atoms at the edges of the layer at large g is:

$$g_0^{(graph)} \approx \frac{2\pi r_0 N}{\langle \mathrm{a} \rangle} = g\frac{2\mathrm{A}_1}{r_0 \langle \mathrm{a} \rangle} \approx \frac{3}{2}\left(\frac{E_{\parallel}}{E_{rad}}\right) \tag{4}$$

where $\langle \mathrm{a} \rangle = \mathrm{a}_{CC}\sqrt{3}$ is the distance between carbon atoms along the zigzag edge of graphene, $\mathrm{A}_1 = (3/4)\mathrm{a}_{CC}^2\sqrt{3}$ is the area per a carbon atom in the graphene layer.

Although (2) and (4) are easy to combine for a graphite cluster, its topology and the growth kinetics are more difficult to be described adequately than that of a diamond. Therefore, with having the goal to propose a method for the synthesis of diamonds (not of graphite), we used the "diamond" ratio (1) in both cases.

2.2. Model of Gibbs Free Energy Change in the Course of Nucleation and Growth of Cluster

In the traditional technique for calculating a new phase nucleation, the total surface energy is substituted into the Gibbs energy change expression ΔG_g of the ensemble of g atoms, when they form cluster that emerges as a result of fluctuations. In the case of the diamond and under the assumption of constant surface energy σ_1 (per one surface carbon atom), the expression for ΔG_g following from (2), has the form (5),

$$\Delta G(g) = \frac{\sigma_1 g}{1+\varpi x} - \varsigma g T \tag{5}$$

while $\varpi = (36\pi)^{-1/3} \approx 0.207$ (for $g \to \infty$ $x \to$ dimensionless cluster radius). It has a maximum at the critical value x_c:

$$\varpi x_c = \left(\sigma_1 - 3\varsigma \mathrm{T} + \sqrt{\sigma_1^2 + 3\varsigma \mathrm{T} \sigma_1}\right) / 3\varsigma \mathrm{T} \tag{6}$$

If in critical nucleus g_c is as small that $\varpi g_c^{1/3} << 1$, then

$$\Delta G(g) \approx \sigma_1 g \left(1 - \varpi g^{1/3}\right) - g\varsigma \mathrm{T} \tag{7}$$

and the parameters

$$x_c = \frac{3(\sigma_1 - \varsigma \mathrm{T})}{4\sigma_1 \varpi}, \quad \Delta G_c = \Delta G(g = g_c) = \frac{27}{256} \frac{(\sigma_1 - \varsigma \mathrm{T})^4}{(\sigma_1 \varpi)^3} = \frac{g_c (\sigma_1 - \varsigma \mathrm{T})}{4} \tag{8}$$

At $\varpi g_c^{1/3} \gg 1$ we have standard expressions of the "drop" model [17-18]. However, at T ~ 0.1 eV and the critical size of $g_c \geq 40$ both terms in the denominator (5) are of the same value. Thus, the Figure 2 shows the dependencies of the position of the maximum x_c and of its value ΔG_c on the supersaturation logarithm ζ according to the "exact" formulas (5), (6) and to the approximate ones (for $\varpi x_c \ll 1$, $\varpi x_c \gg 1$, T = 0.1 eV). In this case, the nucleation barrier becomes zero already at $\varsigma \mathrm{T} \geq 1.2$ (Figure 3). On the contrary, in the drop model, the barrier can only have reasonable value if $\varsigma \mathrm{T} \geq 3$. This value corresponds to giant ζ and to very small nuclei, when the drop model is senseless.

Thus, the correction of the drop model convinces us that the bulky diamond cluster nucleation is possible even at comparatively low supersaturation, i.e., the substrate is an optional element of the problem. For diamond clusters in hydrogen medium, i.e., when the free valences of all surface carbon atoms are saturated with H atoms with a C - H bond energy of ~4.05 eV (i.e., at $\sigma_1 \approx 1.3$ eV), the supersaturation $S_m = \exp(\varsigma_m) = \exp(13)$ provides zeroing of the nucleation barrier even at the temperature of 0.1 eV (1160 K).

T = 0.1 eV. The figure is given within the range of ςT parameter, in which the nucleation barrier has a value of $\Delta G_c \leq 8$ eV, i.e., $\varsigma T \geq 0.6$ (at T = 1000 K this is correspondent with $\varsigma \geq 6$). For the drop model, the value of ΔG_c in the shown range of ςT is very large and is not shown.

In this case, the critical size is several tens of atoms. It only remains to understand whether such values arise in a self-consistent solution.

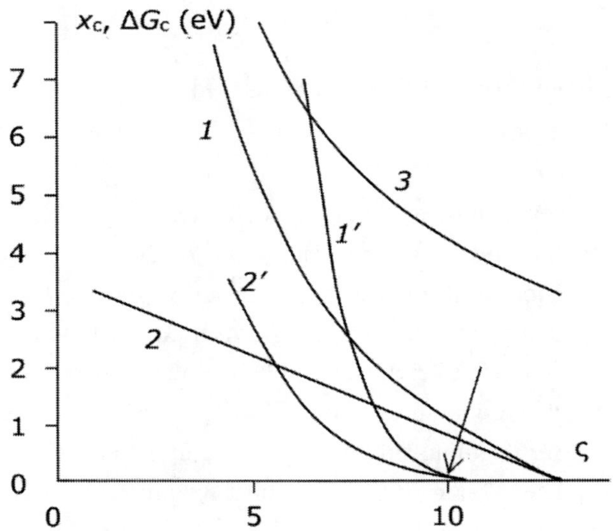

Figure 2. Cluster nucleation free energy maximum value ΔG_c (1,'2,'3') and positions of the maximum (1,2,3) depending on the supersaturation logarithm. 1,1' - according to "exact" formulas (6)-(7); 2,2' - for $\varpi x_c \ll 1$; 3,3' - for $\varpi x_c \gg 1$.

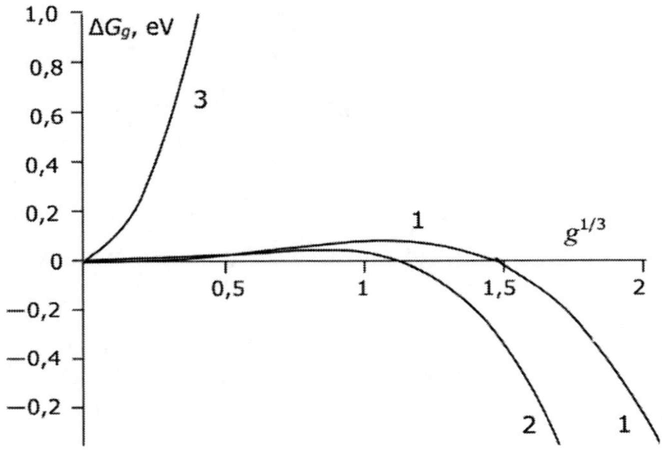

Figure 3. Gibbs free energy change depending on the function $g^{1/3}$. The parameter ςT=1 (arrow in Figure 2). 1 - according to "exact" formulas (6)-(7); 2 - for $\varpi x_c \ll 1$; 3 - for $\varpi x_c \gg 1$ (drop model).

2.3. Comparative Surface Energies of Diamond- and Graphite Clusters

The surface energy σ_1 determines the nucleation barrier ΔG_c, critical for the solid phase nucleation problem. The value of ΔG_c is the sum of the binding energy deficit and the entropy contribution formed by the cluster surface. Within the quantum chemical programs used, the binding energy E_C per carbon atom in an infinite cluster can be estimated as the total binding energy decrease in the course of extraction of a C-atom from the central region of the cluster (without optimizing the geometry of the cluster with arisen vacancy). Then, for g>100 the E_C in the diamond is approximately 16.5 eV (distributed between 4 bonds) and 17.6 eV in the graphite (between three bonds).

The bonding energy deficit connected with the surface can be estimated as $\left(g_C E_C/2 - E_{tot}(g)\right)/g_0$, where $E_{tot}(g)$ is the total binding energy of the cluster with hydrogenated surface. For both the diamond and the graphite clusters, this energy deficit is at the level of hundredths of eV and its sign is unstable. The surface entropy contribution is more significant. To estimate it, the average vibrational frequencies in the bulk of the cluster ν_b and on the surface ν_{surf} were assumed to be independent of the cluster size, and the average vibration frequency of the cluster $\bar{\nu}(g)$ was calculated in the HyperChem package. Herewith, the values of g_C, g_0 for the cluster are known. By comparing two large clusters with the known average vibration frequencies, we reconstructed the surface frequency ν_{surf} and the entropy in the assumption of $\left(\frac{h\bar{\nu}(g)}{T}\right)^{3g-6} \leftrightarrow \left(\frac{h\nu_b}{T}\right)^{3(g-g_0)-6}\left(\frac{h\nu_{surf}}{T}\right)^{3g_0-6}$.

At the temperature of 1000 K, the value of σ_1 per a surface carbon atom increased from 1.0 ± 0.1 eV, 0.9 ± 0.1 eV, for the diamond and for the graphene, respectively, at $g \sim 10$, to 1.25 ± 0.1 eV, 1.2 ± 0.2 eV at $g \sim (50 \div 60)$, reaching saturation. The date scatter for the diamond is associated with different faceting of the clusters; the shape of the graphite clusters was chosen close to spherical to have the contribution of external graphene planes minimal. Thus, the diamond clusters have higher surface energy than the graphite- and the graphene clusters for all *g*. In what follows, the value of σ_1 was assumed to be independent of temperature.

2.4. Desorption of Small Hydrocarbon Species from Surface of Diamond- and Graphene Clusters and Their Formation from Initial Gas Mixture

The carbon cluster nucleation is determined by the balance between the small hydrocarbon species supply onto the cluster surface and desorption of the same species from this surface. In our problem, the first of these factors (the nucleation source) is formed dynamically from the initial gas mixture at rapidly changing temperature. Therefore, to determine the nucleation dynamics, not only the total characteristic reaction heats for the nucleation source formation and for the desorption reactions (that is sufficient in the equilibrium problem) should be known, but the entire sequence of intermediate steps for each of these reactions, i.e., the length of the reaction sequences and the bottlenecks.

The process of desorption is characterized in Figure 4. It shows the changes in the total binding energy of the cluster-vapour system versus the number of the desorbed small species Δn. The system contained a large C_nH_k diamond cluster with a locally broken step (inset in Figure 4, the atom $C_{(2)}$), the desorbed species (CH, CH_3, or C_2H_4), and hydrogen atoms saturating the free valences of carbon atoms on the cluster surface.

The desorption of the species from the diamond step edge starts with breaking one of the C–C bonds of the angular $C_{(2)}$ atom. The formed free valences are partially compensated by adsorption of hydrogen atoms and an additional energy is spent herewith for the molecular hydrogen dissociation. For desorption of the carbon atom $C_{(1)}$ following the $C_{(2)}$ atom, three C–C bonds should be broken. After desorption of two carbon atoms, the topology of the step edge is restored. As it is seen from the inset in Figure 4, the most natural forms of the desorbed species are indeed CH_3, CH_2, CH or $C_2H_{4\text{-}6}$.

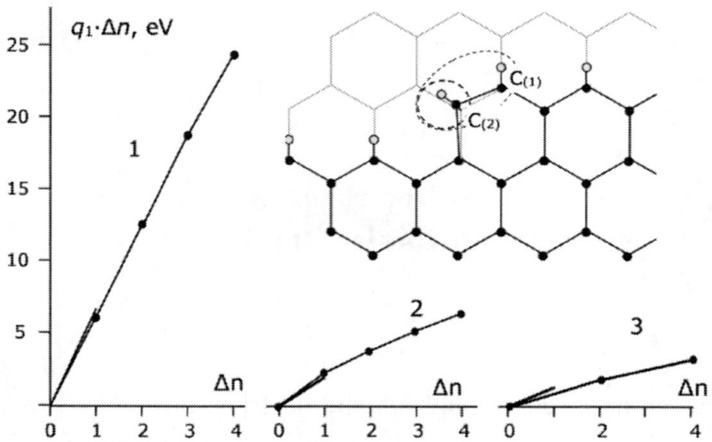

Figure 4. Changes of the total binding energy of the reaction products during the species desorption from diamond cluster: 1 - by reaction $C_nH_k + (1/2)H_2 \leftrightarrow (C_{n-1}H_k) + CH$; 2 - by reaction $C_nH_m + (3/2)H_2 \leftrightarrow (C_{n-1}H_k) + CH_3$; 3 - by reaction $C_nH_k + H_2 \leftrightarrow C_{n-1}H_k + (1/2)C_2H_4$.

For all the reactions, the bold segment shows the slope as the necessary generation energy of the source molecules from methane and hydrogen, calculated per a carbon atom. The horizontal axis is the number of carbon atoms Δn, desorbed from the upper step in the inset.

The inset shows the evaporated step edge of the diamond cluster with carbon atoms $C_{(2)}$, $C_{(1)}$, desorbed one after another within the radicals CH_3 or together with the $C_2H_{4\text{-}5}$ radical. To be short, the atom $C_{(2)}$ is pulled out as related to the lower layer.

The topo-chemical sense of the first steps in the desorption reaction process is shown in more detail in Figure 5. The CH_3 radical desorption from a diamond cluster is taken as an example. The step-by-step sense of these and of the subsequent steps are written out in the Supplement 1.

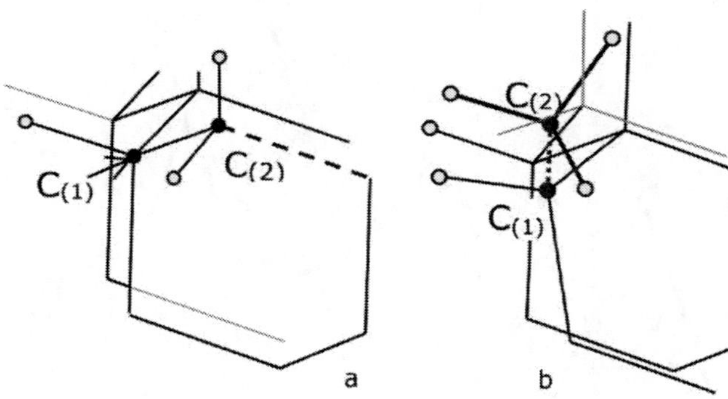

Figure 5. Topochemical sense of the first two steps in the sequence of the desorption reaction of the methyl radical. The final result consists in desorption of two carbon atoms with two CH_3 radicals (see the text). a - the starting configuration, the C-C bond breakdown (dotted line) and the addition of two H-atoms to the ends of the dotted line; b - detachment of the radical $C_{(2)}$-H_3 (highlighted by the bold lines, the broken bond is shown by the dotted line).

The Figure 6 shows the total binding energy change dynamics for all participants of the desorption reaction at intermediate steps for the diamond- and graphene clusters. The final point of the graph corresponds to the desorption of two CH_3 radicals in the reaction $2C_nH_k + 3H_2 \leftrightarrow 2C_{n-2}H_k + 2CH_3$. It s seen that the main contribution to the effective activation energy $q_{btn}^{(des)}$ of the reaction sequence gives the splitting of hydrogen molecules (the "btn" subscription emphasizes the bottleneck of the reaction sequence; the letter "q" is used herewith for the desorption from the cluster, the letter "*E*" for the gas phase source formation).

The reactions of breaking the C–C bonds on the cluster surface are less noticeable, because the vacated valences are compensated by hydrogen atoms. Therefore, the difference between the values of $q_{\mathrm{btn}}^{(\mathrm{des})}$ for the diamond- and graphite clusters are small: $\Delta q_{\mathrm{btn}}^{(\mathrm{des})} \approx 0.6\ \mathrm{eV}$. The difference emerges from the fact that four C–C bonds must be destroyed to desorb two carbon atoms from the diamond cluster against three C-C bonds destroyed in graphite cluster, respectively.

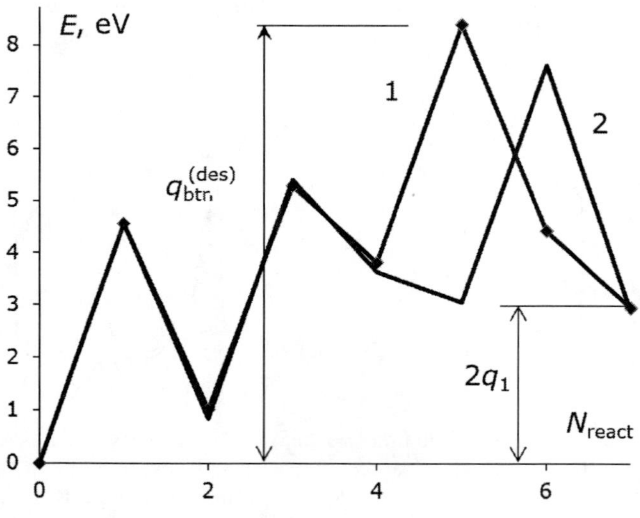

Figure 6. Total binding energy change of the system, including: the diamond (or the graphene) cluster, the reagents and the products involved in stepwise desorption of two methyl radicals from the clusters (see the text). 1 - desorption from the diamond; 2 - desorption from the graphene (graphite).

In Figure 7, a similar sequence of reactions is built for desorption of ethylene C_2H_4. As the carbon dimer of the atoms $C_{(1)}$ and $C_{(2)}$ is completely desorbed now, the dissociation energy of one H_2 molecule is saved, so that the bottleneck of the desorption reaction sequence becomes lower: $\Delta q_{\mathrm{btn}}^{(\mathrm{des})} \approx 5.4\ \mathrm{eV}$ for the diamond cluster exceeds the same value for a graphene cluster by no more than 0.1 eV, i.e., within the scatter in the starting isomer choice. Thus, the desorption of ethylene C_2H_4 can be taken as the main desorption channel from the surface of the clusters.

It should be noted that in the reaction sequences, analysed above, the breakdown of both the C-C- bonds in the cluster and the H-H bonds in the H_2 molecule, as well as bonding of the H-atom with carbon radicals, formed on the diamond surface, are considered as various consecutive reactions. Herewith the first two reactions have high activation energy.

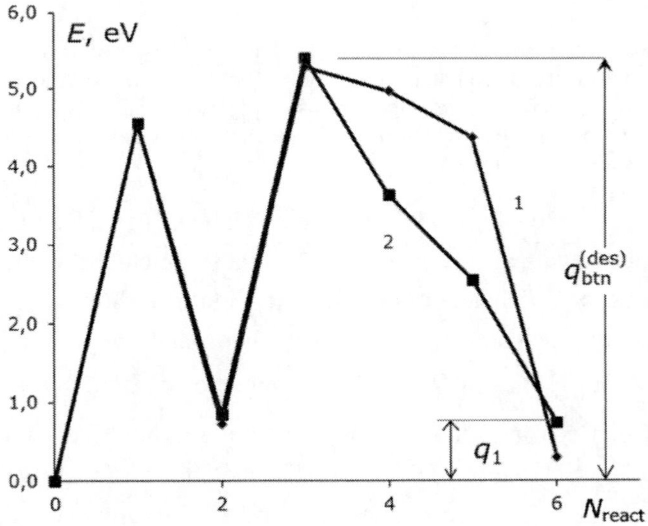

Figure 7. Total binding energy change of the system, including: the diamond (or the graphene) cluster, the reagents and the products involved in desorption of ethylene molecule (see the text). 1 - desorption from the diamond; 2 - desorption from the graphene (graphite).

However, a hypothetical intermediate coordinate-type structure, formed during the formation of the C-H-H-C bridge before breaking the C-C-bond (Figure 8), has larger binding energy value than with a sequential course of events. Such structures can provide smaller activation barrier of the reaction sequence as a whole. Therefore, when evaluating final results, the parameter $\Delta q_{btn}^{(des)}$ is useful to be considered as a variable parameter.

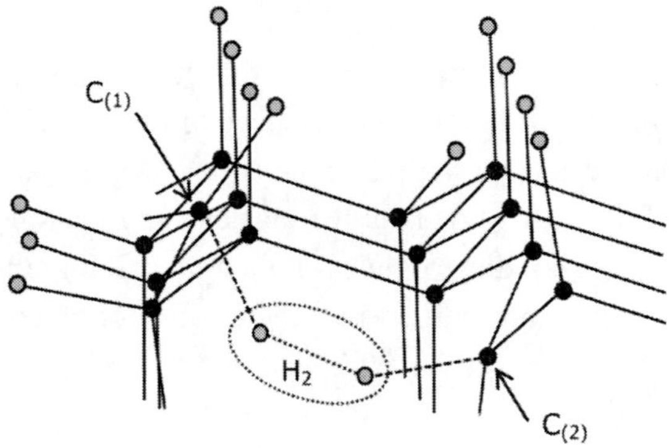

Figure 8. Possible intermediate state of a complex reaction, in which the breakdown of the bond between surface carbon atoms (e.g., atoms $C_{(1)}$ and $C_{(2)}$), dissociation of an H_2 molecule, and fixing the formed hydrogen atoms on the carbon radicals $C_{(1)}$ and $C_{(2)}$ occur as a single process with a significantly smaller activation barrier. The atoms $C_{(1)}$ and $C_{(2)}$ are the same as in Figure 5.

At the same time, the desorbed species CH_3, CH_2, CH or $C_2H_{4\text{-}6}$ are natural for activated gas mixture, which initially includes CH_4, H_2 and oxygen (in one or another compound). The Figure 9 shows the the total binding energy change of small hydrocarbons in reaction $CH_4 \leftrightarrow (1/n)C_nH_m + [(4n-m)/2]H_2$. The number of carbon atoms n in the molecules is plotted along the horizontal axis, the points to the left of n=2 correspond to the CH_3, CH_2, CH radicals. The methane molecule was chosen as the zero-energy point. The Figure 10 shows the binding energy changes per one carbon atom.

There we can see that:

- CH_2- and the CH molecules can be ignored as the nucleation factor;
- the flat graphite is the most favorable cluster in terms of the energy per atom among small clusters in homologous rows of the diamond, the planar (unsaturated) and non-planar (saturated) graphite,
- the diamond and the saturated hydrocarbons of the graphite series are approximately equivalent;

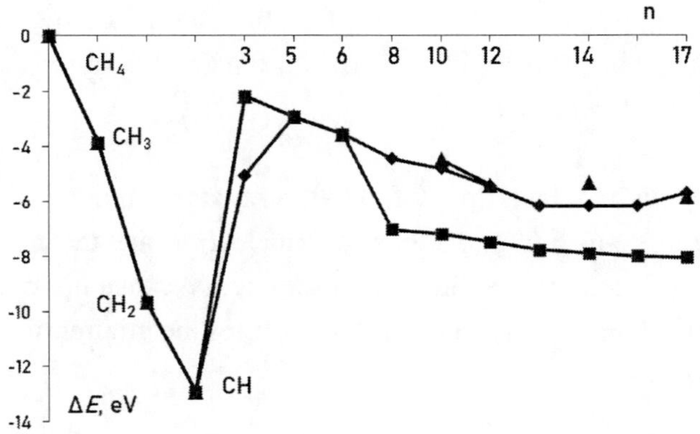

Figure 9. Total binding energy change in the reaction $CH_4 = (1/n)C_nH_m+[(4n-m)/2]H_2$. The symbols ■, ◆, ▼ correspond to flat graphene, graphene with a saturated edge, and diamond, respectively.

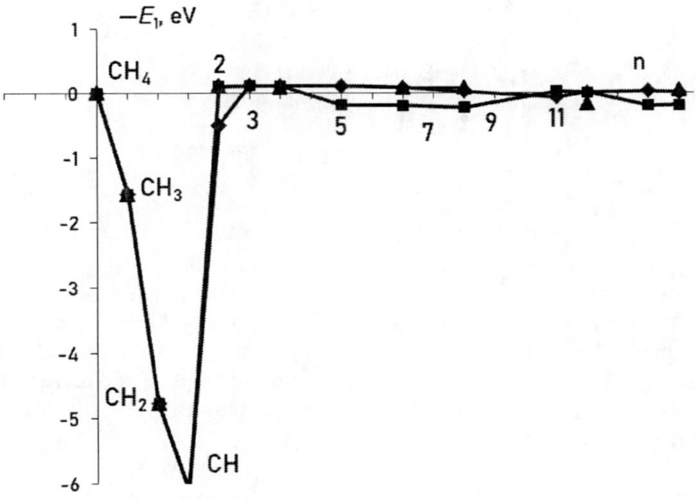

Figure 10. Binding energy change per a carbon atom in the reaction $CH_4=(1/n)C_nH_m+[(4n-m)/2]H_2$. The symbols ■, ◆, ▼ correspond to flat graphene, graphene with a saturated edge, and diamond, respectively.

- in the course of the formation of diamonds, of the graphite-type saturated hydrocarbons, and of planar graphene, there is practically no change in the binding energy and the energies of all bonds are almost constant: the energy of C–H bonds is 4.05 eV (as in methane), $E_{C\text{-}C}$ = 3.88 eV for the graphite-type saturated hydrocarbons, $E_{C\text{-}C}$= (4.5–5.0) eV for the graphite-type unsaturated hydrocarbons.

Therefore, with a fast flow of the gas mixture and with unsteady equilibrium, the main building blocks of nucleation are the methyl CH_3 radicals, which formation requires the least energy consumption, and the ethyl radicals C_2H_4-C_2H_6, which formation requires the minimum number of binary collisions.

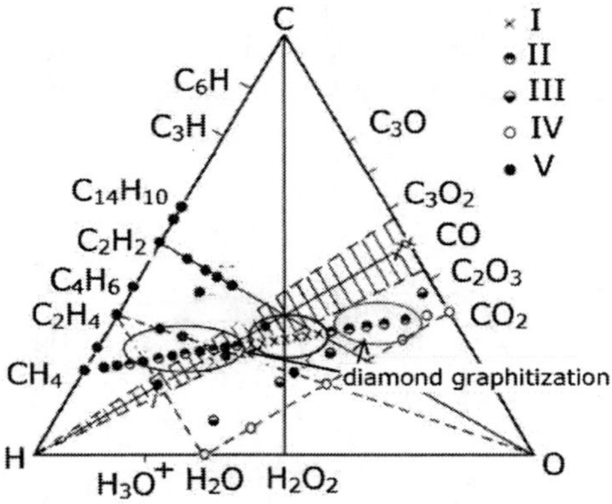

Figure 11. Stoichiometric conditions for the diamond and the graphite poly condensation (borrowed from [5] and significantly simplified): I (central cross marked region) - the diamond growth on the substrate, II - graphitization of diamond (with using the mixture $CH_4 + CO_2$), III - selective oxidation of graphite and preservation of diamond, IV - complete oxidation of both graphite and diamond, V - organics (without details).

The effective activation barrier $E_{btn}^{(diss)}$ for the methane dissociation on the way $H_2 \rightarrow 2H$, $2CH_4 + 2H \leftrightarrow 2CH_3$ is very high (higher than 8 eV). However, it is known that the diamond CVD synthesis is impossible without a significant amount of oxygen. This is seen from the well-known Bachmann diagram (Figure 11) and partly follows from the fact that the graphite surface is oxidized more intensively than the diamond surface. On the other hand, for the nucleation source formation, oxygen (or water vapour) offers a mechanism for detachment of hydrogen atoms from a CH_4 molecule with much smaller barrier $E_{\text{btn}}^{(\text{diss})} \approx (5.5-6)\,\text{eV}$, so that $E_{\text{btn}}^{(\text{diss})}$ and $q_{\text{btn}}^{(\text{des})}$ are approximately equal. Therefore $E_{\text{btn}}^{(\text{diss})}$ can be considered as the bottleneck in the nucleation source formation as a whole.

The necessary dissociation energies of the CH_3, CH, C_2H_4 species according to formal reactions with the initial methane are shown in Figure 4 in the form of tangent bold segments of the curves 1, 2, 3 together with the desorption energies of the same species. It is seen that for all these species the difference E_1—q_1>0 (the factors E_1 and q_1 are defined as positive ones), but small as compared to E_1 and q_1. Therefore, with very slow temperature rise, the diamond nucleation is hardly possible. It is obvious, then, that the sequence of the desorption reactions is much longer than that of the nucleation source formation. With fast increase in temperature, this factor is just as important as the ratio of the bottleneck values of these sequences. Therefore, the supersaturation becomes very high at the very beginning of the picture described in Section I. In addition, it has been shown that it is under these conditions that diamond nucleation occurs much more efficiently.

2.5. Oxidation of Diamond- and of Graphene Clusters

The Figure 12 shows the oxidation energies of the diamond- and graphite surfaces. The fact that the graphite is oxidized much more intensively, is illustrated by the same initial clusters on which the desorption of hydrocarbon species was considered. The general equation of the oxidation reaction with water vapour was $C_nH_m + 2H_2O \leftrightarrow C_{n-2}H_m + 2CO + 2H_2$.

The first steps in the sequence of reactions, shown in Figure 12 and associated with addition of the first water molecule, are described in Supplement 2. It is seen that the effective oxidation barrier of the diamond exceeds that of the graphene (graphite) by 0.7–0.8 eV. Approximately the same difference gives the oxidation reaction with pure oxygen. For what follows, it is most important that the difference in the oxidation barriers is noticeably higher than the difference in the desorption bottleneck energies:

$\Delta q_{\text{btn}}^{(\text{ox})} > \Delta q_{\text{btn}}^{(\text{des})}$. This is due to the fact that after capturing the surface carbon atom by an oxygen atom, the electron density on the C−C bonds in the near-surface region of the cluster shifts towards this atom.

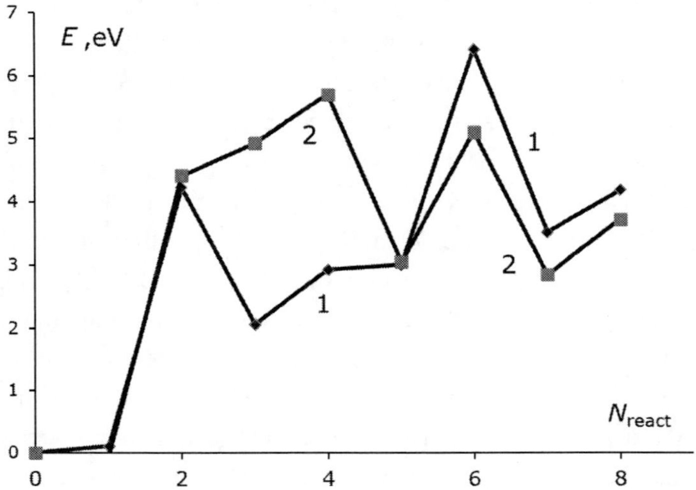

Figure 12. Total binding energy change of the system, including: the diamond (or the graphene) cluster, the reagents and the products involved in sequential oxidation of the clusters with two water vapor molecules (see the text). 1 - oxidation of the diamond cluster; 2 - oxidation of the graphene (graphite) cluster.

The ionic component of C−C bonds appears, and its breaking requires additional energy consumption (even for one additional C−C bond in the diamond).

2.6. Desorption Dynamics Delay of Small Hydrocarbon Species as Compared with Temperature Growth

As the small species desorption from the clusters is a much more multistage reaction that the nucleation source formation, it does not keep pace with the temperature rise.

The equation which describes how the source density $N_1^{(\text{source})}$ approaches the equilibrium density $N_1^{(\text{source,eq})}\left(\mathrm{T}(t)\right)$ at an instantaneous current temperature T(t), has the form:

$$\dot{y}_1 + y_1 \frac{d \ln N_1^{(\text{source,eq})}}{dt} = w_{01} \frac{N}{N_1^{(\text{source,eq})}} \left(1 - y_1\right) \tag{9}$$

where $y_1 = N_1^{(\text{source})} / N_1^{(\text{source,eq})}\left(\mathrm{T}(t)\right)$ is a relative density, $w_{01} = \left(\mathrm{T}/h\right)\left(\Omega_{\text{btn}}/\Omega_0\right)\exp\left(-E_{\text{btn}}^{(\text{diss})}/\mathrm{T}\right)$ is a methane dissociation rate within the transition state theory, $N \gg N_1^{(\text{source,eq})}$ is the total density of carbon atoms in methane and in the source, $N_1^{(\text{source,eq})}\left(\mathrm{T}(t)\right) = \left(\Omega_1/\Omega_0\right)\exp\left(-E_1/\mathrm{T}\right)$, $E_1 \approx 1.6$ eV is the binding energy difference between the start- and the final products of the reaction, synthesizing methyl radical from methane and hydrogen (per a radical CH_3), $E_{\text{btn}}^{(\text{diss})} > E_1$ is the activation barrier of the source synthesis, Ω are the oscillatory and rotational components of partition functions of the original product (methane), of the transition state, and of the nucleation source.

The methyl radical generation is assumed in (9) to be a one-stage process of the methane dissociation, and the activation energy $E_{\text{btn}}^{(\text{diss})}$ is the only bottleneck of the process. If the radical appears as a result of the methane interaction with oxygen or water vapour, which "rips out" a hydrogen atom from the CH_4 molecule, then, according to the HyperChem calculation, the energy $E_{\text{btn}}^{(\text{diss})}$ does not exceed 5 eV.

The exact solution to (9) is not difficult, but in the case of a multistage reaction of desorption, considered below, an analogue of such solution as a solution to a partial differential equation, is possible only numerically. Therefore, the decision was built by interpolating between limiting cases.

In the case of

$$\dot{y}_1 << y_1 \left(d \ln N_1^{(\text{source,eq})} / dt \right) \tag{10}$$

which is correspondent with comparatively high temperature, (9) has a solution:

$$y_1 = \left(1 + \frac{E_1}{T^3} h \dot{T} \frac{\Omega_1}{\Omega_{\text{btn}}} \exp\left(\frac{E_{\text{btn}}^{(\text{diss})} - E_1}{T} \right) \right)^{-1} \approx 1 - \frac{E_1}{T^3} h \dot{T} \frac{\Omega_1}{\Omega_{\text{btn}}} \exp\left(\frac{E_{\text{btn}}^{(\text{diss})} - E_1}{T} \right) \tag{11}$$

in the opposite case:

$$y_1 \approx 1 - \exp\left(-\frac{T^3}{\left(E_{\text{btn}}^{(\text{diss})} - E_1 \right) h \dot{T}} \frac{\Omega_{\text{btn}}}{\Omega_1} \exp\left(-\frac{E_{\text{btn}}^{(\text{diss})} - E_1}{T} \right) \right) \tag{12}$$

The value $N_1^{(\text{source,eq})}(T)$ is estimated by the law of mass action:

$$N_1^{(\text{source,eq})} \approx N_{H_2}^{1/2} N_{CH_4} \left(\frac{h^2}{2\pi m_{H_2} T} \right)^{3/4} \left(\frac{m_{CH_3}}{m_{CH_4}} \right)^{3/2} \upsilon' \exp\left(-\frac{E_1}{T} \right) \tag{13}$$

where $\upsilon' \sim 1$ is the ratio of partition functions of the right- and left side participants in the reaction equation. The factor υ' is a weak function of the temperature.

The process, reverse to the carbon cluster nucleation from their source vapour, is the small species desorption from these clusters. The activation barrier $q_{\text{btn}}^{(\text{des})}$ in the reaction sequence describing this process for the diamond clusters, takes into account the need for the H_2 molecule dissociation to atomic hydrogen and filling the free valences of the carbon surface with these atoms. As it was said above, the value of $q_{\text{btn}}^{(\text{des})}$ is about 5.5 eV in both the diamond- and the graphite channels.

Obtaining solution to a multistage reaction rate under the conditions of fast-changing temperature requires a great deal of simplification.

An approximate expression describing the dynamics of the desorbed species density in the case (10) and derived in Supplement 3, has the form:

$$N_1^{(\mathrm{des})} / N_1^{(\mathrm{des,eq})} \approx \exp\left(-\frac{h\dot{\mathrm{T}}}{\mathrm{T}^2} \int_0^{\mathrm{s_{btn}}} \frac{q_\mathrm{s}^{(\mathrm{des})}}{(d/ds)\left(q_\mathrm{s}^{(\mathrm{des})} + q_\mathrm{a,s}^{(\mathrm{des})}\right)} \exp\left(\frac{q_\mathrm{a,s}^{(\mathrm{des})}}{\mathrm{T}}\right) ds \right)$$

where $N_1^{(\mathrm{des,eq})}(\mathrm{T}(t))$ is the equilibrium density of desorbed species in assumption of constant concentration of methane, the integration variable "s" is the number of the reaction stage, $q_\mathrm{a,s}^{(\mathrm{des})}$ is the activation energy of stage "s," $q_\mathrm{s}^{(\mathrm{des})}$ is the total binding energy of the products at the stage "s," counted from the similar value of the initial reagents of the desorption process.

As this energy $q_\mathrm{a,s}^{(\mathrm{des})}$ is connected with transition state calculation, its determination is not as correct as that of the stationary state energies $q_\mathrm{s}^{(\mathrm{des})}$. Therefore, we estimated it for qualitative consideration by the value of $q_\mathrm{btn}^{(\mathrm{des})}$ or simply supposed it to be a variable parameter. If $q_\mathrm{s}^{(\mathrm{des})}$ depends linearly on s and the activation energy is constant ($q_\mathrm{a,s}^{(\mathrm{des})} = q_\mathrm{a}^{(\mathrm{des})}$), then

$$N_1^{(\mathrm{des})} / N_1^{(\mathrm{des,eq})} \approx \exp\left(-\frac{h\dot{\mathrm{T}}}{\mathrm{T}^2} \frac{\mathrm{s_{btn}^2}}{2} \exp\left(\frac{q_\mathrm{a}^{(\mathrm{des})}}{\mathrm{T}}\right) \right) \tag{14}$$

A similar expression in the case, opposite to (10) (low temperature), is given in the Supplement 3.

When evaluating the equilibrium density of species $N_1^{(\mathrm{des,eq})}(\mathrm{T})$, we will consider (as in Section II.4), that desorption from the diamond- or from the graphene cluster $N_1^{(\mathrm{des,eq})}(\mathrm{T})$ proceeds by reaction

$C_nH_k + 2H_2 \leftrightarrow C_{n-2}H_k + C_2H_4$, where n and k are the numbers of carbon- and hydrogen atoms in the cluster. As the densities of the clusters C_nH_k and $C_{n-2}H_k$ are close, the mass action law gives

$$N_1^{(\text{des,eq})} \approx N_{H_2}^2 \left(h^2 / 2\pi m_{H_2} \mathrm{T}\right)^{3/2} \upsilon \exp\left(-q_1 / \mathrm{T}\right) \tag{15}$$

where $q_1 = E_{C_2H_4} - 2E_{H_2} + E_{C_{n-2}H_k} - E_{C_nH_k}$ is the total binding energy change of the components participating the desorption of ethylene (dominating among the desorbed hydrocarbon species). The meaning of the constant υ is the same as of υ' in (11).

2.7. Estimation of Nucleation Onset Moment at Supersaturation Decay of Small Hydrocarbon Species

It follows from (11) - (15) that the supersaturation S increases exponentially with the value of $\dot{\mathrm{T}}$. Herewith the logarithm $\zeta = \ln S$

$$\varsigma \approx \left[\ln \frac{N_{H_2}}{N_{CH_4}} + \frac{1}{\mathrm{T}}\left(q_1 - E_1\right) + \frac{h\dot{\mathrm{T}}}{\mathrm{T}^3}\left(\frac{\mathrm{Ts_m}^2}{2} \exp\left(\frac{q_a^{(\text{des})}}{\mathrm{T}}\right) - E_1 \exp\left(\frac{E_a - E_1}{\mathrm{T}}\right)\right)\right] \tag{16}$$

increases drastically with increase in T. On the contrary, its negative derivation $\left(d\varsigma / dt\right)_{\text{nat}}$, connected with natural change of ζ, decreases drastically by its absolute value.

Let us determine the moment when the natural decay rate of the supersaturation logarithm $\left(d\zeta / d\mathrm{T}\right)_{\text{nat}}$ on the curves ζ(T) in (15) turns out to be less by absolute value than the nucleation decay rate. It is known [18, 20] that at high and relatively low supersaturation, this rate should be described in different ways. In the first case, the nucleation occurs practically without a barrier; in the second, the explosive nucleation rate is proportional to $\exp\left(-\Delta G_c^{(m)}/\mathrm{T}\right)$ with the critical nucleus formation barrier $\Delta G_c^{(m)}$ at a previously unknown maximum supersaturation; to determine it, we should solve the explosive nucleation problem [21-23].

In our case, the saturated vapour density over the cluster surface is formed with a long delay; the nucleation develops from extremely high supersaturations, so that the analysis should be started with the barrier-free case. Then the ζ decay rate during the nucleation is

$$\left(d\varsigma/dt\right)_{\mathrm{nucl}} = \dot{\mathrm{T}}\left(\varsigma/N_1^{(\mathrm{source})}\right)\left(dN_1^{(\mathrm{source})}/d\mathrm{T}\right) \tag{17}$$

We can estimate the decrease in the small species density (the nucleation source decay $dN_1^{(\mathrm{source})}/dt$) at the expense of nucleation by the dimerization rate

$$dN_1^{(\mathrm{source})}/d\mathrm{T} = -\left(\sigma_{\mathrm{crs}}\upsilon_{\mathrm{T}}/\dot{\mathrm{T}}\right)\left(N_1^{(\mathrm{source})}\right)^2 \exp\left(-E_{\mathrm{rep}}/\mathrm{T}\right) \tag{18}$$

The index 1 means a conditional "monomer," the cross-section σ_{crs} index for collision of the species emphasizes the difference from the surface tension (also denoted by σ), the energy E_{rep} is the necessary energy for overcoming repulsion when the species merge.

With taking (10) into account, it is easy to show then that the temperature $\mathrm{T}_{\mathrm{melt}}$, intersectional for the curves $\left|(d\varsigma/d\mathrm{T})_{\mathrm{nat}}\right|$ and $\left|(d\varsigma/d\mathrm{T})_{\mathrm{nucl}}\right|$, depends on $q_{\mathrm{a}}^{(\mathrm{des})}$ logarithmically. Its value is ~1500 K with $\dot{\mathrm{T}} = 1000\,\mathrm{K/s}$ and the densities of $N_{\mathrm{CH_4}} = 100$ torr, $N_{\mathrm{H_2}} = 1$ atm. The meaning of this result is that condensation cannot start until the temperature reaches a certain value.

As the graphite and the diamond have approximately values of $q_{\mathrm{a}}^{(\mathrm{des})}$, nucleation process can start in the both channels. If $q_{\mathrm{btn}}^{(\mathrm{eff})}$ for the diamond is yet higher than for the graphite, the conditions for nucleation in the graphite channel occur earlier. However, if nucleation of the graphite clusters is impossible for some reason, the diamond cluster nucleation occurs almost instantaneously. The reason for the impossibility can be the presence of oxygen. In the MWCVD- and the HFCVD- syntheses, it is the presence of oxygen that ensures the choice of the diamond channel rather than of the graphite channel. This is due to a fact that more stable diamond nuclei are oxidized to a much smaller extent than the graphite ones. This fact can also be seen in the Bachmann diagram showing that a significant amount of oxygen is absolutely necessary for the synthesis of diamond.

Therefore, it is in the diamond channel that the carbon cluster nucleation is to be awaited to occur. However, the more correct estimate should consider the possibility of nucleation at any current temperature and supersaturation logarithm $\zeta(\mathrm{T})$, assuming that these values correspond some metastable state. The fictitious history of how these conditions have arisen could lie to the left of the moment t_{m} in Figure 13, is unimportant for our problem and marked by dotted line.

Thus, the problem of a metastable state relaxation with starting parameters $\zeta_{\mathrm{m}}, t_{\mathrm{m}}, \mathrm{T}_{\mathrm{m}}$, is being solved, in which the values $\zeta_{\mathrm{m}}, t_{\mathrm{m}}, \mathrm{T}_{\mathrm{m}}$, are associated with the current values ζ, t, T in the actual process. The explosive (expl.) stage of the nucleation ends at the moment when the supersaturation decay rate $\left|(dS/d\mathrm{T})_{\mathrm{expl}}\right|$ reaches its maximum and starts to decrease (the lower arrow in Figure 13 and the arrow in Figure 14).

In the semi-analytical solution to the “classical” nucleation problem [21-23], this rate is found under the assumption of quadratic decay law for $\varsigma(\mathrm{T})$ beyond the maximum ζ_{m}. Then the decay rate of ζ can be considered maximum at the moment when the actual density of the source N_1 decreases as related to $N_1^{(\mathrm{source})}$ by “e” times: $\Delta N_1/N_1^{(\mathrm{source})} \approx 1-1/e$ (Figure 14). If the explosive stage is too slow in comparison with the value of $\left|(d\varsigma/d\mathrm{T})_{\mathrm{nat}}\right|$, the nucleation is impossible with the given law of the $\varsigma(\mathrm{T})$ variation. On the contrary, if the half-width of the nucleation curve is small (as in Figure 13), the nucleation is determined by the moment of intersection of the curves $\left|(d\varsigma/d\mathrm{T})_{\mathrm{nat}}\right|(\mathrm{T})$ and $\left|(d\varsigma/d\mathrm{T})_{\mathrm{expl}}\right|(\mathrm{T})$.

Like the "classical" nucleation problem [21-23], the process is calculated from the balance equation for the substance (carbon) in the carbon species and in their nucleation products (the diamond or graphite clusters): the decrease in the species density N_1 relative to $N_1^{(\mathrm{source})}$ is equal to the number of condensed species ΔN_1:

$$N_1^{(\mathrm{source})} - N_1 = \Delta N_1 = \int_0^t J(\tau)\, g(\tau \to t)\, d\tau \tag{19}$$

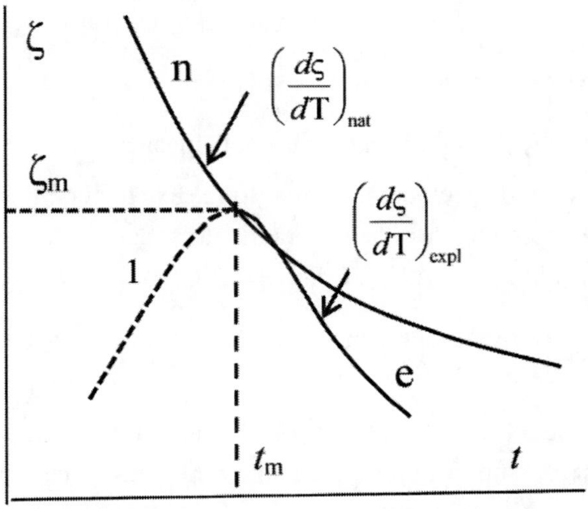

Figure 13. Schematic for estimating the explosive nucleation temperature $\mathrm{T}_{\mathrm{nucl}} = \dot{\mathrm{T}} t_{\mathrm{m}}$ in the course of natural supersaturation decay caused by the temperature increase. The curves "n" and "e" are the natural and explosive (i.e., caused by explosive nucleation) supersaturation logarithm ζ decay rates. The dotted line 1 is a fictitious prehistory of the metastable state creation with a given supersaturation logarithm ζ_m at the moment t_m.

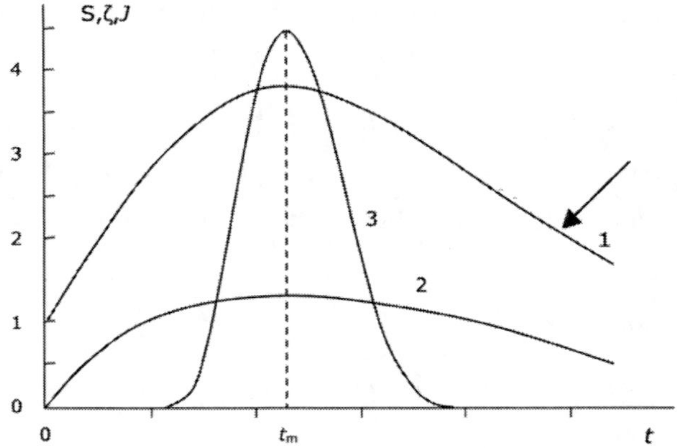

Figure 14. Hierarchy of characteristic times put in the model of the explosive nucleation stage [6-7]. Curves 1 and 2 are S and $\zeta=\ln S$, respectively, J is the flow density of the clusters through the bottleneck in the size space. The curve 3 is much narrower than 1,2. The scale of the flow is arbitrary. The arrow marks the moment of the maximum supersaturation decay rate.

The value of N_1 is composed of the small species CH_n, C_2H_m densities, the collisions and rearrangement of which form the clusters. For simplicity, we assume that the nucleation source density change is insignificant and describe the cluster flow $J[\mathrm{cm}^{-3}\mathrm{s}^{-1}]$ through the size space bottleneck $(x_c, \Delta G_c)$ in (19) by one of the versions of the Zeldovich's model [17, 18].

Within this model, we use the quasi-stationary approach, assuming that the flow tracks the supersaturation change, and the species spatial flow density onto a growing cluster is determined by the Knudsen expression $\frac{1}{4} N_1 \upsilon_T$:

$$J^{-1} = (1/N_1)\sum_{g=1}^{\infty}\left(w_g\Omega_g\right)^{-1} \approx \left(\pi N_1^2\upsilon_T\right)^{-1}(4\pi c/3g_c)^{2/3}\int_1^{\infty} dg/\Omega_g$$

$$J \approx \frac{N_1^2\upsilon_T}{6Sc}\left(\frac{4\pi c}{3}\right)^{1/3}\exp\left(-\frac{\Delta G_c}{T}\right)\sqrt{\frac{2\pi T\varpi\sigma_1}{(\sigma_1-\varsigma T)^2}} \tag{20}$$

where "*c*" is the carbon atoms density in the condensed phase (in diamond), υ_T the thermal velocity of the species. The Courtney factor (supersaturation *S*) is added to the denominator of (21) "by hand." The second factor in the integrand in (19) is the growth rate of the number of atoms *g* in the cluster:

$$\frac{dg}{dt} = 4\pi r_g^2 c \frac{dr_g}{dt} = N_1 \upsilon_T \pi r_g^2 \tag{21}$$

whence it follows that the cluster radius r_g grows as

$$r_g = \frac{N_1 \upsilon_T t}{4c} + r_c, \text{ and with } r_g \gg r_c: \qquad g = \frac{4}{3}\pi c \left(\frac{N_1 \upsilon_T t}{4c} \right)^3 = \frac{\pi N_1^3 \upsilon_T^3 t^3}{48c^2}.$$

Thus, both factors in the integrand in (19) are determined. The technique of reducing this equation (19) to an algebraic equation system in the solution, obtained in [21-23], consists in single- and then double differentiation of (19) in time and the closure of the system by equilibrium relations for the source vapour density. In our case, these procedures are not necessary. The initial supersaturation and temperature are the parameters, and the integrand in (19) can be easily written. So, the difference between N_1 and $N_1^{(\text{source,eq})}$, caused by the cluster nucleation and by transition of a part of the source into the clusters by the moment *t* is

$$J(\tau) g(\tau,t) = \theta \sum_{p=0}^{5} \left[\begin{array}{l} C_5^p \left(-\frac{E_1}{\mathrm{T}^3} h \dot{\mathrm{T}} \frac{\Omega_1}{\Omega_{\mathrm{btn}}} \right)^p \cdot \\ \exp\left(\left[p\left(E_{\mathrm{btn}}^{(\mathrm{diss})} - E_1 \right) - \left(\Delta G_c + 5E_1 \right) \right] / \mathrm{T} - \varsigma \right) \end{array} \right] (t-\tau)^3 \tag{22}$$

where $\Delta G_c = \sigma_1 x_c^4 / 4\mathrm{T}$, C_5^p is a factorial multiplier in the Newton binomial theorem,

$$\theta = \frac{\pi \upsilon_{\mathrm{T}}^{4}}{288 c^{3}} \left(\frac{4\pi c}{3} \right)^{1/3} \sqrt{\frac{2\pi \mathrm{T} \varpi \sigma_1}{(\sigma_1 - \varsigma \mathrm{T})^2}} \left[N_{\mathrm{H}_2}^{1/2} N_{\mathrm{CH}_4} \left(\frac{h^2}{2\pi m_{\mathrm{H}_2} \mathrm{T}} \right)^{3/4} \left(\frac{m_{\mathrm{CH}_3}}{m_{\mathrm{CH}_4}} \right)^{3/2} \upsilon' \right]^5 \tag{23}$$

The formula (22) considers that the nucleation temperature, obtained further, is comparatively high, so that the value of ς in (20) is calculated from (9). All five components of the sum in (22) have the structure

$$\theta C_5^p \left(-\frac{E_1}{\mathrm{T}^3} h \dot{\mathrm{T}} \frac{\Omega_1}{\Omega_{\mathrm{btn}}} \right)^p \exp(-f_p)(t-\tau)^3$$

where

$$f_p = \left[(\Delta G_{\mathrm{c}} + 5E_1) - p\left(E_{\mathrm{btn}}^{(\mathrm{diss})} - E_1 \right) \right] / \mathrm{T} + \varsigma \tag{24}$$

is a drastically changing function of ς. Since the structure of the components is the same, we give for brevity the result of the solution to (19), which takes into account only one of them (p =0): the maximum decay rate of the supersaturation logarithm beyond the nucleation maximum is

$$\left(\frac{d\varsigma}{d\mathrm{T}} \right)_{\mathrm{expl}} = -\left(\dot{\mathrm{T}} \right)^{-1} \sqrt{\frac{8}{g_{\mathrm{c}}^{(\mathrm{m})}} \left(\frac{\theta}{2\Delta N_1} \right)^{1/2}} \exp\left(-f_0^{(\mathrm{m})} / 2 \right) \tag{25}$$

where $f_0^{(\mathrm{m})} = \varsigma_{\mathrm{m}} + \Delta G_c^{(\mathrm{m})} / \mathrm{T}_{\mathrm{m}}$, $\Delta G_c^{(\mathrm{m})} = \sigma_1 \varpi \left(x_c^{(\mathrm{m})} \right)^4 / 4$, the upper- or the lower symbol «m» is correspondent with the critical parameters at the supersaturation maximum moment. The result (25) is obtained in Supplement 3.

Comparison of typical curves $\left| (d\varsigma / d\mathrm{T})_{\mathrm{nat}} \right|$ and $\left| (d\varsigma / d\mathrm{T})_{\mathrm{expl}} \right|$ (Figure 15) shows that the condition $\left| (d\varsigma / d\mathrm{T})_{\mathrm{nat}} \right| \geq \left| (d\varsigma / d\mathrm{T})_{\mathrm{expl}} \right|$ for both the graphite and the diamond is valid within a temperature range of no more than several tens of K.

Herewith, the temperature dependence of the interception point of the curves $|(d\varsigma/d\mathrm{T})_{nat}|$ and $|(d\varsigma/d\mathrm{T})_{expl}|$ on the temperature rise rate $\dot{\mathrm{T}}$ is weak. Thus, the curves (almost the lines, to be more correct) shown in Figure 16, have very different values of $\dot{\mathrm{T}}$: $\dot{\mathrm{T}} = 800\ \mathrm{K/s}$ and $\dot{\mathrm{T}} = 1300\ \mathrm{K/s}$ (but is should be noted that the model is initially built for high values of $\dot{\mathrm{T}}$ and the results cannot be extrapolated to $\dot{\mathrm{T}} \to 0$).

On the contrary, as it follows from Figure 16, the moment of the nucleation onset is strongly dependent of the activation energy $q_{a,s}^{(des)}, \mathrm{s} \to \mathrm{btn}$ at the final stages of desorption. As already said, for qualitative consideration the value of $q_{a,s}^{(des)}$ can be estimated by the value of $q_{s}^{(des)}, \mathrm{s} \to \mathrm{btn}$ or simply be supposed a variable parameter.

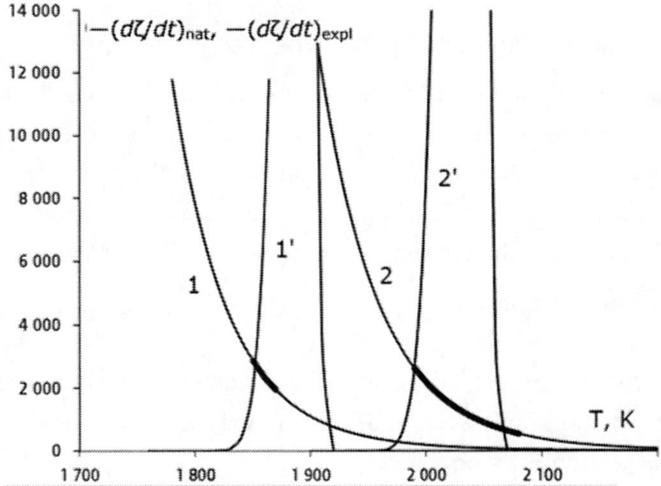

Figure 15. Supersaturation logarithm decay rate in the channel of the graphite (1.1’) and of the diamond (2,2’). $q_{btn}^{(des)} = 5.4\ \mathrm{eV}$. 1,2 - natural decay due to the temperature rise, 1,’ 2’ - decay caused by explosive nucleation of clusters.

It was obtained in Sections II.4-II.5 that the difference of the values of $q_{\mathrm{btn}}^{(\mathrm{des})}$ in the diamond- and graphite channels lies within the scatter of initial clusters and the methods of their calculation. Let us assume, however, that the diamond has $q_{\mathrm{btn}}^{(\mathrm{des})}$ slightly higher than the graphite has: $q_{\mathrm{btn}}^{(\mathrm{des,diam})} = 6.4\ \mathrm{eV}$, $q_{\mathrm{btn}}^{(\mathrm{des,graph})} = 6.0\ \mathrm{eV}$.

Then the nucleation conditions in the graphite channel are achieved at lower temperature, i.e., at an earlier point in time, and nucleation of the graphite clusters should be expected. However, the nonoxidation requirement for small clusters establishes an opposite relationship between the synthesis of the graphite- and diamond clusters. It can be illustrated by estimating the growth rate of the carbon atom number in a cluster (at least at the level of the drop model) by the equation:

$$\frac{dg}{d\mathrm{T}} \sim \left(N_1 \upsilon_{\mathrm{T}}^{(1)} \sigma_{\mathrm{crs}}^{(1)} - N_{\mathrm{ox}} \upsilon_{\mathrm{T}}^{(\mathrm{ox})} \sigma_{\mathrm{crs}}^{(\mathrm{ox})} \exp\left(-E_{\mathrm{ox}} / \mathrm{T}\right) \right) g^{2/3} \tag{26}$$

where N_{ox} is the oxidant density, $\upsilon_{\mathrm{T}}^{(1)}, \sigma_{\mathrm{crs}}^{(1)}, \upsilon_{\mathrm{T}}^{(\mathrm{ox})}, \sigma_{\mathrm{crs}}^{(\mathrm{ox})}$ are the thermal velocities and the collision gas-kinetic cross-sections of the cluster with the species and the oxidant, respectively. E_{ox} is the activation barrier in the oxidation reaction of the carbon cluster, accompanied by appearance of a CO_x-type fragment, the first term contains the source density N_1, i.e., it is proportional to exp(ζ). The carbon cluster will grow if the right side of (26) is positive. This is true if $\varsigma \geq \ln \frac{N_{\mathrm{ox}}}{N_{\mathrm{CH_4}}} - \frac{E_{\mathrm{ox}} - E_{\mathrm{btn}}^{(\mathrm{diss})} - E_1}{\mathrm{T}} + \mathrm{Const}$. On the "graphite" curve in Figure 15, it is satisfied on a very short section of temperature variation (bold section of the curve 1), on the "diamond" curve (curve 2) this section is much wider.

The meaning of this result is as the follows: the difference between the effective oxidation energies in the diamond- and the graphite channels $E_{\mathrm{ox}}^{(\mathrm{diam})} - E_{\mathrm{ox}}^{(\mathrm{graph})}$ is much larger than the difference between the desorption bottlenecks of small hydrocarbon species $q_{\mathrm{btn}}^{(\mathrm{des,diam})} - q_{\mathrm{btn}}^{(\mathrm{des,graph})}$. Thus, the oxidation energy of the graphite clusters $E_{ox}^{(graph)}$ in (26) is significantly lower than that of the diamond clusters, and their appearance requires higher supersaturation. The oxidation energy of the diamond clusters is much lower, so that permissible supersaturation for them can also be lower. In fact, it turns out to be higher than in the graphite channel (Figure 17), and the emerging diamond clusters are viable.

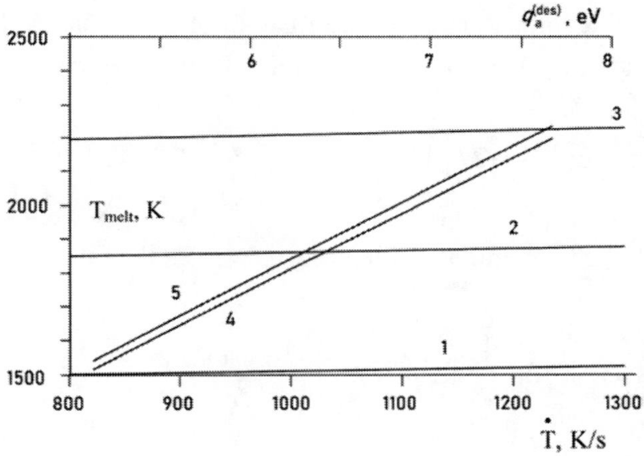

Figure 16. Cluster nucleation temperature in the diamond channel, depending on: 1,2,3 - the temperature rise rate $\dot{T}$; $\dot{T}$ changes along the lower abscissa axis. $q_a^{(des)}[eV]$: 5.1(1), 6.4(2), 7.7(3); 4,5 - the activation energy of final steps of the desorption process $q_a^{(des)}$. $q_a^{(des)}$ changes along the upper abscissa axis. $\dot{T}[K/s]$: 800 (4), 1300 (**5**).

Thus, when considering the carbon cluster synthesis with fast increase in temperature, the synthesis of diamond cluster is more advantageous than that of the graphite clusters.

The model, initially developed for the case of fast increase in T, indicates a slight increase in the nucleation temperature T_{cross} with an increase in the derivative $\dot{T}$ (Figure 16, the derivative $\dot{T}$ is plotted along the lower abscissa axis). With increase in $\dot{T}$ (i.e., with an "excessively" fast increase in T), the diamond cluster density $N_g^{(diam)}$ and the "contrast factor"

$N_{\bar{g}}^{(\mathrm{diam})} / N_{\bar{g}}^{(\mathrm{graph})}$ also decrease. The both of the values $N_{\bar{g}}^{(\mathrm{diam})}$ and $N_{\bar{g}}^{(\mathrm{graph})}$ can be determined at the same level of an estimate as (26):

$$\frac{dN_{\bar{g}}}{d\mathrm{T}} = \mathrm{a}S^2 \exp\left(-2E_1 / \mathrm{T}\right) - \mathrm{b}N_{\mathrm{Ox}} N_{\bar{g}} \exp\left(-E_{\mathrm{ox}}^{(\mathrm{diam})} / \mathrm{T}\right) \tag{27}$$

where "a" and "b" are the universal constants (but different, of course) for the graphite and diamond. Then, at the final stage of the nucleation, when the left-hand side of (31) vanishes, the diamond cluster density

$$N_{\bar{g}} \sim \exp\left(\left(E_{\mathrm{ox}}^{(\mathrm{diam})} - 2E_1\right) / \mathrm{T}\right), \tag{28}$$

decreases with the temperature T increase, as well as the ratio

$$\frac{N_{\bar{g}}^{(\mathrm{diam})}}{N_{\bar{g}}^{(\mathrm{graph})}} \sim \frac{S_{\mathrm{diam}}^2}{S_{\mathrm{graph}}^2} \exp\left(\frac{E_{\mathrm{ox}}^{(\mathrm{diam})} - E_{\mathrm{ox}}^{(\mathrm{graph})}}{\mathrm{T}}\right), \quad E_{\mathrm{ox}}^{(\mathrm{diam})} > E_{\mathrm{ox}}^{(\mathrm{graph})}.$$

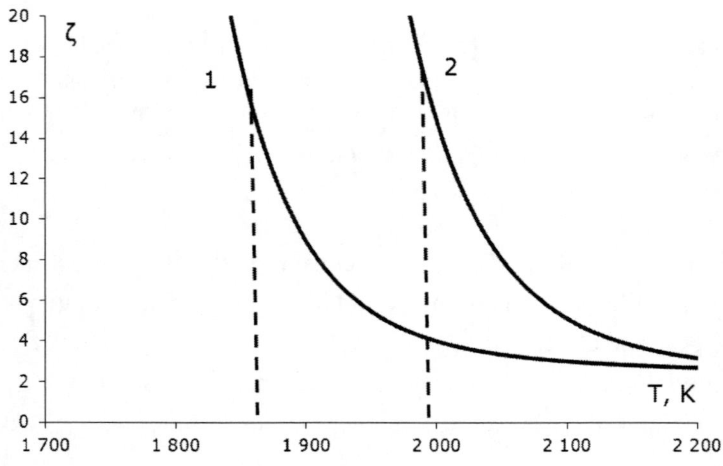

Figure 17. Changes of the value of $\zeta = \ln S$ in the graphite (1) and in the diamond (2) channels with the temperature rise (the nucleation start moment is marked by dotted line for the both channels). The start of the diamond cluster nucleation corresponds to a slightly higher supersaturation.

Due to the condition (27) the number of atoms *g* in the diamond cluster also becomes smaller when T_{cross} increases as the clusters grow less intensively and oxidize faster.

The dependence of the nucleation temperature T_{cross} on the bottleneck energy in the desorption reaction sequence $q_{btn}^{(des)}$ is much stronger (curves 4, 5). After the cluster nucleation, which occurs within the temperature range of several tens of K, these clusters can only remain intact if the temperature increase is changed by decrease towards the substrate. This transition in both the HFCVD and the MWCVD techniques does indeed occur. In MWCVD, it is achieved by the fact that the electric field oscillation node (where the field is maximum), is always located at a certain distance from the substrate.

The estimate of the carbon cluster nucleation temperature of 1800–1900 K, obtained in the self-consistent solution, clearly overestimated in comparison with the substrate temperature T_s and gas temperature T_g, typical of the MWCVD process. However, for the HFCVD process, it matches the temperature of the filament well. As to the typical conditions of the MWCVD diamond synthesis, activated by microwave discharge plasma in a fast gas flow, there is no time for a common temperature to be established for the heavy component T_g, electrons T_e, and vibrations T_v in the carbon clusters. At the same time, different components of the carbon cluster synthesis are sensitive to different temperatures.

An effective temperature T_{eff} for the small clusters and for hydrogen in the equilibrium distributions and reaction probabilities can be estimated as the geometric mean of T_g and T_e. If the preexponent in the direct reaction of the H_2 molecule dissociation is taken from the numerical calculation results in isothermal hydrogen plasma [24] and an independent relationship between T_g and T_e is known, we can find these values separately. We used the pure hydrogen plasma calculation results that simulated the MWCVD gas dynamics in the diamond synthesis geometry [25, 26].

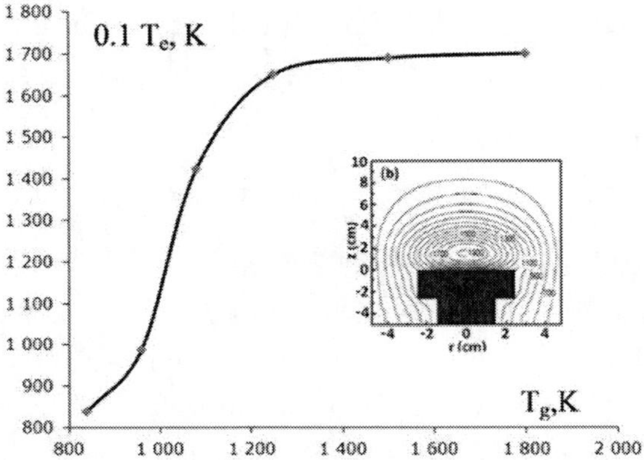

Figure 18. Electron temperature T_e depending on the gas temperature T_g in the gas - plasma flow around the substrate, calculated in [26].

The distribution $T_e(\mathbf{r})$ from [26] is shown in the inset in Figure 18. The Figure 18 as a whole restores the dependence $T_e(T)$ from distributions $T_g(\mathbf{r})$ and $T_e(\mathbf{r})$. Although the calculation [26] is only valid for a certain microwave power, flow rate and geometry, we assumed the dynamics of the ratio $(T_e—T_g)$ to be universal. Then, at the nucleation temperature of ~1900 K, the gas temperature lies below 1000 K. As it should be, this value is somewhat higher than the substrate temperature typical of the diamond synthesis.

Then the estimates show that T_{eff} found above corresponds to the relationship between T_g and T_{eff} if T_{eff} is understood as the characteristic vibrational temperature of clusters T_v. Indeed, the difference between T_v and T_g is determined by non-isothermal nucleation effects in small clusters. The value of T_v is higher than that of T_g due to adsorption from vapour and desorption heat release, while the temperature is equalized due to inelastic collisions of the cluster with hydrogen molecules. Then the difference between T_g and T_v can be estimated as

$$3g\frac{d}{dt}\left(\mathrm{T_v}-\mathrm{T_g}\right)=(1/4)\mathrm{A_g}\upsilon_{\mathrm{T}}^{\mathrm{CH_3}}\left(N_1q_1\left(1-S^{-1}\right)-N_{\mathrm{H_2}}\sqrt{\frac{m_{\mathrm{H_2}}}{m_{\mathrm{CH_3}}}}\frac{2m_{\mathrm{H_2}}}{m_{\mathrm{g}}}\left(\mathrm{T_v}-\mathrm{T_g}\right)\right),$$

where $\upsilon_{\mathrm{T}}^{(\mathrm{CH_3})}$ is the thermal velocity of methyl radicals, A_g is the cluster surface area, estimated as the ratio of the surface cluster atom number $g_0(g)$ to the area of nonplanar hexagonal cell projected onto the (111) plane: $A_6=(3/2)a_{CC}^2\sqrt{3}$. Equating the logarithmic derivatives T_v-T_g and T_g, at the estimation level, we obtain

$$\frac{\mathrm{T_v}-\mathrm{T_g}}{q_1}\approx\left(\mathrm{A_g}\,(1/4)\upsilon_{\mathrm{T}}^{(\mathrm{CH_3})}N_1\right)\left(3g\frac{\mathrm{T}'}{\mathrm{T}}+\frac{1}{4}\mathrm{A_g}\left(\frac{m_{\mathrm{H_2}}}{m_{\mathrm{CH_3}}}\right)^{3/2}\upsilon_{\mathrm{T}}^{(\mathrm{CH_3})}N_{\mathrm{H_2}}\frac{2m_{\mathrm{CH_3}}}{m_{\mathrm{g}}}\right)^{-1}.$$

In different modes, T_g is 1.5–2 times smaller than T_v.

The substrate does not participate in the above consideration directly. The diamond clusters arisen in the gas phase are deposited on the substrate already formed. However, the factors that determine the fast-heating criterion play their role in the CVD diamond synthesis technique on any substrate. Therefore, considering the ideology of the CVD synthesis for massive acceptable quality diamonds as fundamentally exhausted, we can apply this criterion to the other possibilities of obtaining diamonds. The idea of control the gas-phase condensation channels by non-stationary temperature dynamics can work under the conditions of the synthesis not on a substrate, but in the bulk of a medium, which limits free growth of the diamond.

3. Nucleation and Growth of Diamond Clusters in Carbonate-Silicate Matrix

It is known that unconstrained expansion of the diamond material even on a prepared single-crystal diamond substrate does not allow high-quality diamond to be obtained in the MWCVD technology. A natural idea to solve this problem is to constrain free growth of the diamond material in different crystallographic directions. So, it is proposed to grow the diamond clusters in a template that would compensate for the excessive surface energy. Besides that, the template should be mobile, easily deformed during the diamond growth, and prevent from the diamond cluster dissolution.

A natural candidate for the role of such template is a silicate-carbonate melt. The cluster placed within this melt is forced to grow with a counter pressure from the melt, maximal in the direction where the diamond tends to grow most rapidly. Still more important is the fact that the carbon material supply to the diamond crystal through a chemically inert matrix is much slower than in the CVD- process.

In a similar way, the growth of a longer silicon carbide polytype SiC (instead of a shorter SiC polytype) occurs with a slower supply of the "building material" and with time sufficient for relaxation of the growing lattice.

The diamond nucleation and the growth in silicate-carbonate melt (or in another melt) is not a new idea, but a traditional trend in the diamond synthesis. However, the process is usually carried out under ultra-high pressures in various types of the HPHT setup. The melt contains a diamond seed and graphite material which is transformed into the diamond. Such a parameter as the temperature rise rate is not used in calculations and is not reflected in theoretical models. However, the fact that such models have not yet optimized composition of the melt reflects, to our opinion, fundamental problems of the HPHT technique and the need for other approaches. We believe that the use of the dT/dt factor let make an advance in technology of massive SCD diamonds production without using the pressure which significantly exceeds atmospheric pressure.

To our opinion, the diamond synthesis in the melts without ultrahigh pressure can be organized as the follows. The mineral mixture is placed horizontally across the gas-phase carbon carrier flow (e.g., $CO + CH_4 + H_2$), by flat layer in a gas-permeable crucible, hermetically adjacent to the walls of a pumping reactor, placed inside an inductor. The mixture is brought into melt phase by means of induction heating. With a pressure drop applied across the melt layer, the gas mixture passes through it as gas bubbles of the same composition.

A portion of the methane diffused into the melt is dissociated to small gas-phase species CH_3, C_2H_4, etc.

If the melt contains the carbonate component, an internal gas-phase CO_2 is unintendedly added, associated with decomposition of the low-melting carbonate component according to the scheme $MeCO_3 \rightarrow MeO + CO_2$ (Me is a bivalent metal). Since this moment, the solid phase diamond synthesis in the melt can be considered like the gas-phase synthesis in Section II, assuming that supersaturation with carbon, initiating nucleation, is determined with respect to the gas phase. The only difference is that carbon atoms are delivered to the growing cluster by diffusion through the melt cells.

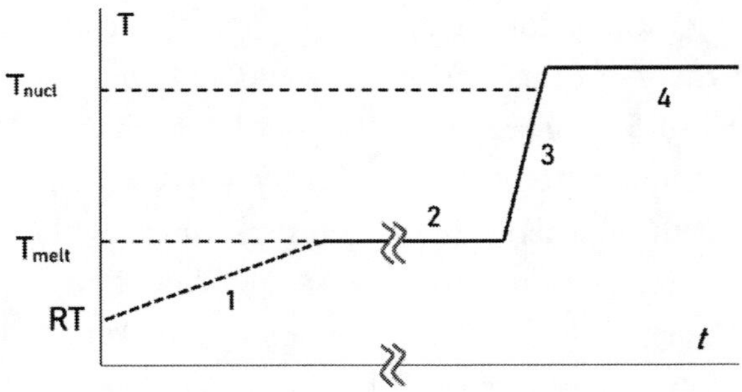

Figure 19. Proposed temperature dynamics in a carbonate-silicate melt layer for nucleation and growth of diamond clusters: 1 - crucible content heating up to a temperature not lower than the melting point T_{melt}; 2 - melt saturation by a gas-phase carbon carrier; 3 - rapid-heating ended by the formation of diamond clusters; 4 - Ostwald ripening of the clusters and their subsequent growth; The ratio of durations of the temperature change steps is arbitrary.

The main building blocks of the diamond clusters (like the MWCVD case), are CH_3 and C_2H_n species ($n \leq 5$). On the one hand, their density in the gas phase is maximum compared to the densities of the other small hydrocarbon species. On the other hand, the carbon species desorbed from a growing diamond cluster have the same shape. This is true regardless of whether the evaporated carbon atom is on a singular surface (e.g., (111)) or on the edge of a step: in both the cases, it is bonded to three nearest carbon atoms and with one hydrogen atom, which compensates for the missing valence.

In the presence of an extra gas-phase source (CO) an additional atomic carbon source appears and, apparently, the density of the CH, CH_2 species increases. Herewith, this density as the density of the desorbed saturated vapour increases with a significant delay.

The fast temperature rise forms the diamond clusters in the melt by the same mechanism as in the CVD. Then the temperature is supposed to be stabilized and the clusters grow slowly in the heterogeneous catalytic reactor mode with a slight non-equilibrium (Figure 19). The necessary components of the proposed technique that must be verified by a simulation and estimates are as follows.

1. The melt should envelop the cluster in such a way that the protruding carbon atoms would be saturated with silicon atoms; the binding energy of this saturation would be high to compensate the excessive surface energy of the cluster better than hydrogen atoms do. The compensation factor should be manifested much stronger as related to the cluster diamond clusters, than to the graphite (or graphene) ones, i.e., the diamond cluster should be more profitable from the very beginning than the graphite one. At the same time, the oxygen atoms of matrices of the SiO_2, CO_3 type should not be in direct contact with the diamond cluster body;
2. With a reasonable difference in the pressure between the inlet and outlet of the reactor with the melt layer, the gas flow rate through the melt must be high enough for the quasi-cells of the melt to be saturated with carbon for an acceptable time; for this, the melt quasi-cell size must be large enough;
3. The small gas-phase species density must ensure the supersaturation of the silicate or of the carbonate phase with carbon.
4. The small species should be brought to the cluster and removed from it at a rate that tracks the temperature rise dynamics.

5. The carbon cluster interaction with the matrix should not form foreign solid- or gaseous phases, silicon carbide and free silicon, first of all.
6. The areas of the controlled temperature rise must exist inside the melt, where the temperature tracks the power variation of the induction heating. In these areas, the conditions for both the diamond cluster origin at fast-heating stage and their further growth at constant temperature stage must be ensured.
7. After the explosive nucleation at the stage of the temperature rise is completed and temperature is constant, the clusters must not be dissolved.

The reality of technical conditions that satisfy these demands (and the other, arising from a more detailed consideration) is evaluated below.

3.1. Simulation of How Growing Graphite- or Diamond Clusters Interact with Melt. Selection of Melt

The interaction of the growing graphite- or diamond clusters with the melt was analysed, as before, in the HyperChem package. We began the consideration with the melts of SiO_2, $Mg(Ca)CO_3$, in which the diamond clusters nucleation looks the most natural. When stimulating the most thermally steady of mineral components (SiO_2) we imitated its structure with a cubic silicon lattice, in which an oxygen atom was inserted into each Si-Si bond (Figure 20).

This structure does not correspond to any real SiO_2 polymorph, but the difference in polymorphs is levelled in the melt. Anyway, the oxygen atoms with a lone electron pair are placed between the nearest Si atoms not in a straight line, but with an obtuse angle. As the direction of this angle under melt conditions is stochastic, the resulting quasilattice resembles cubic or hexagonal lattice very vaguely and can imitate the properties of the melt.

Similarly, when simulating the quasilattice of the $Mg(Ca)CO_3$ melt, we proceeded from a hexagonal crystal lattice in which double layers, similar to the (111) planes of the diamond lattice, contain carbon atoms. The pairs of divalent Mg(or Ca) atoms are located between these layers, the oxygen atoms are located between each pair of the carbon atoms (Figure 21).

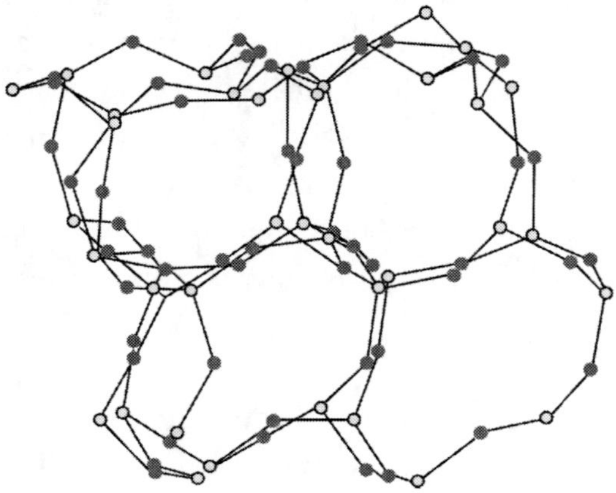

Figure 20. Quasi-lattice of SiO_2 melt, used in the diamond nucleation modelling in this melt. The white and grey circles represent silicon and oxygen atoms, respectively. To be short, the hydrogen atoms, present in the original structure in HyperChem for calculating the thermodynamic properties, are omitted.

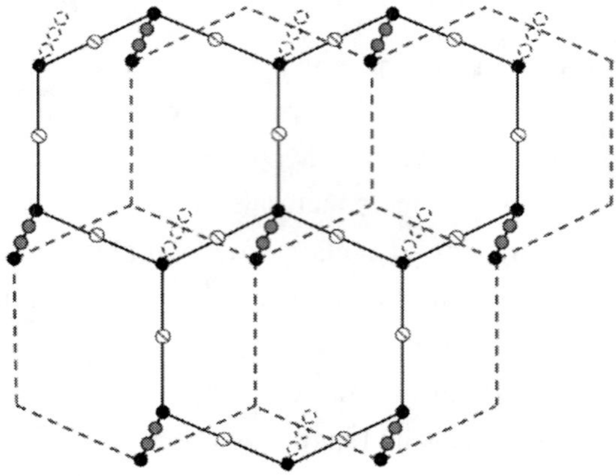

Figure 21. Structure of the carbonate $Mg(Ca)CO_3$ crystal lattice, accepted as the starting structure when modelling carbonate melt by HyperChem tools. Double planes similar to the diamond (111) planes, contain atoms C (black), pairs of bivalent Mg(Ca) atoms placed between the layers (grey), and the atoms "O" (crushed circles) between each pair of carbon atoms. In fact, the detailed representation of the carbonate lattice is not important, because in the diamond cluster nucleation regime, carbonate is either decomposed to MeO and CO_2, or balances between melting and decomposition.

Like the MWCVD- and the HFCVD techniques, both saturation and supersaturation of the melt with carbon is ensured by fast increase in the melt temperature (at the same rate of about 1000 K/s, or at another calculated value), which ensures the small species synthesis. The temperature rise can be provided by a programmed short-term increase in the induction power applied to the melt.

In order to understand the details of the mutual arrangement of the atoms when carbon clusters are formed in the melt (e.g., in SiO_2), a diamond cluster with a size slightly larger than the quartz melt cell, was first built. Then the "protruding" atoms of this cluster, i.e., the atoms with only two or three carbon neighbours were "freed" from the H atoms (which were the elements of an auxiliary structure), and a hypothetical "zero iteration" of the atom arrangement was constructed. Then radii were drawn through the centre- and "protruding" carbon atoms of the cluster, on which the silicon atoms Si were fixed. Their distance to the "protruding" carbon atoms was chosen somewhat longer than the length of the Si-C bonds in SiC. Those Si atoms that were closer to the neighbouring Si atoms than they should be in the quartz melt were removed. Then the Si atoms were "loaded" with four oxygen atoms. The Si-C bonds for the silicon atoms were not fixed and in the course of the calculation had to either appear or be rejected. Similarly, the positions of the oxygen atoms with unfilled valences were not fixed.

It was assumed that the mineral shell, constructed this way, is either "wetting" or "unwettable" as related to the diamond cluster. Simulation indicated the first option, i.e., all the Si–C bonds are reproduced in the expected positions (Figure 22), the cluster-melt structure is determined correctly and energetically favourable. All the oxygen atoms appear farther from the cluster surface than the Si atoms, "finding" other silicon atoms.

To estimate the surface energy compensation of the diamond cluster when it is placed in the mineral shell, it is necessary to know the binding energies Si–C, C–C, C–H and characteristic entropies of these interactions.

The energies of the Si–C bonds in SiC and of the C–C bonds in the diamond were estimated by removing a pair if neighboring atoms in the bulk of the SiC- and the diamond clusters and determining the total binding energy change without optimizing the geometry of disturbed systems. The energies were $E_{Si-C} = 3.45$ eV and $E_{C-C} = 4.0$ eV, respectively, higher than the standard values but in the same proportion. Approximately the same value of E_{Si-C} remained at the interface of the carbon cluster. It was determined by removing the entire diamond cluster from the shell with knowing the energies of both the entire system and the cluster separately.

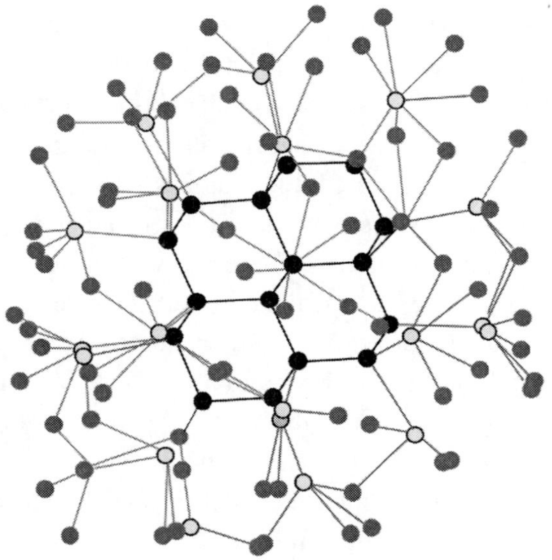

Figure 22. Diamond cluster, surrounded by a SiO_2 shell. The black, white and grey circles represent carbon, silicon and oxygen atoms, respectively. All bonds other than C-C bonds are shown by grey lines.

Unfortunately, it is too difficult to extract the surface entropy by this way. Nevertheless, it is natural to think that with very disordered structure of the cluster-shell interface, this entropy should be higher than the interfacial entropy of the diamond cluster coated with hydrogen atoms, and the surface free energy is lower (by absolute value). Therefore, the replacement of the surface hydrogen atoms to silicon atoms is thermodynamically favourable, as well as the replacement of silicon atoms to carbon atoms, embedded in the diamond structure, in the cluster expansion process.

When comparing the free surface energy compensation due to silicon in the diamond and the graphite clusters, it is essential that the surface carbon atoms in the diamond are located not closer one from another than at the distance of $a_{CC}\sqrt{3}$ (i.e., the inter-atomic distance on the (111) diamond plane). Accordingly, the distance between the Si atoms that form the shell of the diamond cluster turns out to be approximately the same as in the SiO_2 structure and these Si atoms do not repulse one from another when they cover the diamond cluster.

On the contrary, in the graphite cluster carbon atoms at the edge are located closer to each other and have two free valences. When they all are saturated with silicon atoms, the binding energy of the system decreases notably by absolute value (as the simulation in HyperChem shows). Therefore, the most of the valences of the edge carbon atoms must be saturated with hydrogen atoms, not with atoms Si, and compensation function of the silicon is not substantial for the graphite. On the contrary, silicon, interacting with the external graphene planes, can transform the system towards the diamond.

When the carbonate melt interacts with an emerging diamond cluster, the compensation function of Mg- or Ca atoms is substantially weaker and the carbonate component role of the initial mineral matrix is different. In order for the diamond cluster to grow, it must be located in the melt medium from the very beginning. Herewith, the carbonate component provides the melting point decrease of the system as a whole. So, magnesium carbonate melts with decomposition to MgO and CO_2 at 540°C. Potassium carbonate (Potash) melts at 890°C to be decomposed on K_2O and CO_2 at 1200°C. It is also important that with an additional power supply an extra carbon source appears.

The idea of using a combined melt correlates well with experiments in the different silicate-carbonate melts within the HPHT technology. An overview of the literature on this area can be found in [27]. Thus, in purely silicate systems such as H_2O-SiO_2 and SiO_2-C (graphite), the diamond is not crystallized [28]. At the same time, SiO_2 in small values was observed in the melt composition where the diamond was obtained in the mildest conditions. So, in [27, 29], the carbonate-silicate melt included amorphous SiO_2 and carbonates K, Na, Ca, Mg. The spontaneous diamond nucleation minimum parameters did not exceed 5.5 GPa and 1250°C. In the experiment, described in [30], the carbonate component included the K_2CO_3, and the content of SiO_2, necessary for the nucleation, did not exceed 25%. At the same time, it is known that silicate and carbonate melts are completely mixed to form homogeneous melt.

The fact that pure silicate melt cannot be an effective diamond nucleation medium, follows from the chemical reaction analysis of the emerging diamond clusters. In the C + SiO_2 system, such clusters do not differ from carbon coke used in carbothermic synthesis of the silicon carbide [31-32]. The carbon cluster can disappear then by the following reactions:

$$SiO_2^{sol} + C^{(diam)} \leftrightarrow SiO + CO \tag{29}$$

$$SiO_2^{sol} + 2C^{(diam)} \leftrightarrow Si + 2CO \tag{30}$$

$$SiO_2^{sol} + 3C^{(diam)} \leftrightarrow SiC + 2CO \tag{31}$$

each of them being complicated by the presence of hydrogen and has non-trivial topo-chemical sense. Estimates of the binding energies and the entropies according to standard statistical mechanics formulas, using the HyperChem package show the following.

In contrast to (29), (30), the reaction (31) is endothermic, (binding energies are: $E_{diss}(CO) = 9.8$ eV, $E_{diss}(SiC) \sim (5.0 \div 5.3)$ eV, $E_{diss}(C) = 6.9$ eV, $E_{diss}(SiO_2) \sim 9.0$ eV, $\Delta E_{diss} \approx$ —5 eV, the diamond binding energy is determined here as ½ in the binding energy change when the diamond step on the plane (111) is reduced by 2 carbon atoms, insert in Figure 3). The increase in entropy when the gas product is formed in the reaction (31), provides the Gibbs energy decrease at the temperatures above 1900 K only, which is consistent with estimates in [32].

The absolute binding energy increase is $\Delta E_{diss} \sim 1$ eV (if the binding energies are: $E_{diss}(CO) = 9.8$ eV, $E_{diss}(SiO) = 7.1$ eV, $E_{diss}(SiO_2) \approx 9.0$ eV) in the reaction (29), $\Delta E_{diss} > 2.0$ eV in the reaction (30) ($E_{diss}(Si) = 4.61$ eV, the crystalline silicon binding energy was determined in (34) in the same way as for the diamond). Besides that, the gas-phase product synthesis in the reaction (30) and especially in (29) is accompanied by the entropy production. Thus, the reactions (29), (30) are spontaneous and can only be partially blocked by the temperature decrease to at least 1400-1500 K. Correspondence of this temperature to the diamond nucleation temperature is discussed further. Here it is significant that the temperature of 1400 K lies below the melting point of the pure SiO_2.

Herewith, the cluster formation mechanism assumes that the cluster is surrounded by the melt, and not by a solid phase. That means that the SiO_2 shell must be mixed with a substance that reduces the melting point. The role of this substance, able to limit the melting temperature as low as to 800 K, can be played by carbonates. If the optimal temperature rise rate T' is the same as in the MWCVD method in Section II, the interval between 800 K and the nucleation temperature of ~1400 K must be passed for ~1 s, and survival of a part of the clusters should be provided with gas-phase reaction product CO, SiO supply.

It is impossible to create the conditions for so fast heating throughout the whole melt layer thickness. With a standard temperature $T_{melt} + 10$ K, the electrical conductivity of the KCl melt is $\sigma \approx 2.0\ (\Omega{\cdot}\text{cm})^{-1}$. Herewith the skin-layer $\Delta(\text{mm}) = 503/\sqrt{\sigma[\text{SI}]\nu}$ [33] can be as thick as 1 mm (i.e., its thickness is comparable with the melt layer thickness) if only the frequency is $\nu = (1/\sigma[\text{SI}])(503/1)^2 \sim 1\ \text{kHz}$. This frequency is related to mean frequency region, quite typical for the induction heating technique. Herewith, the thermal conductivity of the melt $\kappa = (0.4 \div 1)\ \text{W/(m}\cdot\text{K)} \sim 10^{-2}\text{W/(cm}\cdot\text{K)}$ is still high enough to ensure a modest temperature drop inside the layer.

Indeed, the temperature difference across the melt layer is not difficult to estimate, assuming that the whole flow density ΔP, which enters the melt layer through one of the boundaries, is removed from this boundary by heat radiation outward and inside the melt by the thermal conductivity: $\Delta P = \sigma_{St} T_0^4 + (\kappa / h)|dT/dx|$, where σ_{St} is the Stefan-Boltzmann constant, h is the melt layer thickness, $\left|\frac{dT}{dx}\right|$ is the temperature gradient, normal to layer on its border.

With increase in power, which is deposited inside a thin layer, this gradient is determined by the unsteady thermal conductivity equation inside the layer in the absence of internal heat sources: $\frac{\partial T}{\partial t} = \chi \frac{\partial^2 T}{\partial x^2}$ where $\chi = \frac{\kappa}{\rho c_P}$ is the temperature conductivity, ρ and c_P are the density and heat capacity of the melt. According to $\dot{T} = \chi T_{xx}$, $\left|\frac{dT}{dx}\right| = \frac{T''_{xx} h}{2} = \chi^{-1} \dot{T} h/2$, $\Delta P = \sigma_{St} T_0^4 + \frac{\rho c_P}{2} \dot{T}$, and the temperature difference between the inlet- (and the outlet) melt border and the central plane of the melt layer is $T_0 - T_c = \frac{T''_{xx}}{2}\left(\frac{h}{2}\right)^2 = \left(\frac{\kappa}{\rho c_P}\right)^{-1} \dot{T} h^2/8$.

With the temperature rise rate of 500 K/s, $h = 1$ cm and $\kappa = 1.0 \cdot 10^{-2}$ (W/(cm·K)) the difference $T_0 - T_c$ is almost as high as the temperature itself. Therefore, T' should not be above ~100 K/s, and only in this case $T_0 - T_c = 400$ K is relatively small. The power density ΔP does not exceed (40-50) W/cm^2. For modern RF setup [33] for hardening the metal surfaces this value is small, so that the conditions of the RF induction heating can be realised.

3.2. Conditions for Saturation and Supersaturation of Melt Cells with Carbon Carrier

The time, necessary for saturation of the melt layer with the thickness h by gas-phase carbon, can be estimated as the diffusion time in the capillary system. The transverse size of quartz (or of carbonate, Figure 20, 21) quasi-cell is about $d_{eff} = 0.7$ nm. With the effective transverse movement amplitude in the pore $d_{eff} - d$ (where $d \sim$ 0.3 nm is a characteristic CH_3 molecule diameter), accepted as the mean free path, we obtain the diffusion flow velocity of $v \sim (d_{eff} - d) v_T / 3h \approx 0.4 \times 10^{-7} \times 10^5 / 3 \sim 10^{-3}$ cm/s. Thus, the time interval h/v which is necessary to equalise the species' densities in the layer with $h\sim 1$ cm, is not in excess of an hour (the temperature in the thermal velocity v_T estimate is put 1000 K). The gas-phase pressure difference between the sides of the layer is not involved in this estimate. This difference can enter the viscous Stokes velocity of the gas flow in a cylindrical capillary caused by a longitudinal pressure gradient. It gives about the same result.

The gas transfer through the melt in the form of bubbles does not make a noticeable acceleration of the saturation process. The size of such bubble r can be estimated from the fact that the Laplace pressure change in the bubble can hardly exceed this pressure itself p_0, i.e., $r \sim 2\sigma / p_0$. With the melt surface tension of $2.5\cdot 10^3$ dyn/cm the bubble radius is $r[\text{cm}] \sim 5\cdot 10^{-3} / p_0[\text{atm}]$. So, the ascent velocity, found from equality of the floating force to the Stokes resistance: $\rho g r \cdot \pi r^2 = 6\pi\eta r \text{v}$, is

$$\text{v} = \frac{\rho g r^2}{6\eta} \sim \frac{2\cdot 1000\cdot 25\cdot 10^{-6}}{6\cdot 1[\text{puas}]p_0^{\,2}[\text{atm}]} \sim \frac{10^{-2}}{p_0^{\,2}[\text{atm}]} \qquad (32)$$

So, both the floating velocity and the bubble size can be controlled by the viscosity (which is set by the temperature) and the pressure. Thus, before the start of the cluster nucleation process, the melt must be supported in the gas flow for at least a few hours. Such condition is unlikely to complicate the synthesis significantly. With subsequent fast temperature increase, the gas-phase system is practically isolated.

The fast-heating mechanism for such a system can work if the monomers of the growing cluster are supplied to the cluster and removed from it as fast as to track the temperature change dynamics. For this, the order of magnitude of characteristic monomer supply time which is determined by the reverse flow to the cluster must be smaller than a second.

Feasibility of such a criterion is almost the same as in the case of the Knudsen monomer movement regime in the gas phase. Indeed, the diffusion flow density onto the cluster has an order of magnitude $J_{\text{diff}} \sim D\nabla N_1^{(\text{source})} \sim DN_1^{(\text{source})} / r_{\text{clust}}$. The diffusion constant D in the melt medium is determined by the melt cell size a_{cell} and even at the size of a_{cell}=7 Å is greater than $D \sim (1/3)a_{cell}v_T \sim 2\cdot10^{-2}\,\mathrm{cm}^2/\mathrm{s}$.

With the cluster size $r_{clust} = 3$ Å and the monomers density of $10^{16}\,\mathrm{cm}^{-3}$, the reverse time is $\sim 4\pi D r_{clust} N_1^{(source)} \sim 10^{-8}\,\mathrm{s}$. It almost does not differ from the Knudsen regime estimate when the time has an order of $\left[(4\pi r_{clust}{}^2/4)N_1^{(source)}v_T\right]^{-1} \sim 0.3\cdot10^{-8}\,\mathrm{s}$, despite the difference between the thermal and diffusion velocities. The reason is that the flows in the diffusion- and in the Knudsen regime are proportional to the first- and to the second power of the cluster radius, respectively.

It follows from the above that the temperature of carbon cluster nucleation (unlike the temperature rise rate), does not differ substantially from the temperature T_{melt}, obtained in Section II for the MWCVD and the HFCVD processes. The temperature T_{melt} was determined in Section II by the intersection point of the "natural" supersaturation decay rate $|(d\zeta/dT)_{nat}|$ and the rate $|(d\zeta/dT)_{expl}|$, obtained from the solution to the explosive nucleation problem, where the current temperature and supersaturation are selected as some initial metastable state parameters (Figure 13).

The value $|(d\zeta/dT)_{nat}|$ is related with the temperature rise rate and does not depend on the details of the small carbon species feeding and removing. These details are contained in the value of $|(d\zeta/dT)_{expl}|$, that describes the kinetic or diffusion cluster growth regime.

Like Section II, $|(d\zeta/dT)_{expl}|$ was calculated from the balance of the substance (carbon) in the small species and in the nucleation products (clusters): the decay of the fragment (or monomer) density N_1 as related to the value $N_1^{(source)}$ calculated from the formula (20) is equal to the number of carbon atoms in the condensed species ΔN_1.

3.3. Diamond Cluster Nucleation Problem for Carbonate-Silicate Melt

The difference from (20) is that the methyl radicals are supplied to the cluster and remove from it by diffusion through the melt cells. Herewith, the expressions for both the source flow density in the size space *J* and the number of atoms in the cluster $g(\tau,t)$ during the cluster growth from the starting critical size g_c are modified. The expression for *J* was obtained in [34]:

$$J = \sqrt{\sigma / \mathrm{T}}\left(DN_1^{(\mathrm{des,eq})} / 2\pi c^2 r_c^3\right) N^{(\mathrm{clust})}\left(r_c\right) \tag{33}$$

where the size distribution function of the clusters $N^{(\mathrm{clust})}(r)$ with a sharp minimum near the critical radius r_c is

$$N^{(\mathrm{clust})}(r) \sim N^{(\mathrm{clust})}(r_c) \exp\left(4\pi\sigma\left(r - r_c\right)^2 / \mathrm{T}\right) \tag{34}.$$

The value of $N_1^{(\mathrm{des,eq})}$ is the equilibrium density of the small species desorbed from the clusters surface at temperature T. The surface tension σ, in contrast to Section II, has standard dimension. The formulas (33) - (34) are obtained from the drop model of the cluster. However, in the final expression the exponent power, which is the strongest factor in the nucleation rate, is restored with parameters of the more accurate model, like the Section II, so that the drop model tracks remain only in weak pre-exponential factors.

From normalization (34) to the total small species density and from the fact that in the entire region $r < r_c$ of the cluster radius the value of $N^{(\mathrm{clust})}(r)$ does not differ substantially from the equilibrium function, the flow (33) acquires the form:

$$J(\mathrm{T}(\tau))=(\mathrm{T}/\sigma)^{1/2}\left(D\left(N_1^{(\mathrm{source})}\right)^2/2S\pi^2\varsigma^3\right)\exp(-\Delta G_c/\mathrm{T}) \tag{35}$$

The other multiplier under the integral (19) is given by solving to the problem of the cluster size r increase at $r >> r_0 > r_c$:

$$r=\sqrt{2DN_1^{(\mathrm{des,eq})}\varsigma/c}\,(t-\tau)^{1/2},\ g=(4/3)\pi c\left(2DN_1^{(\mathrm{des,eq})}\varsigma/c\right)^{3/2}(t-\tau)^{3/2} \tag{36}$$

The expressions for the source density and for the equilibrium desorbed species density are given by (11)-(12), (14) in Section II. Then, the integrand in (19) is

$$J(\tau)g(\tau,t)=\theta\sum_{p=0}^{2}C_2^p\left(-\frac{E_1}{\mathrm{T}^3}h\dot{\mathrm{T}}\frac{\Omega_1}{\Omega_{\mathrm{btn}}}\right)^p\cdot\exp(-f_p)(t-\tau)^{3/2} \tag{37}$$

where

$$\theta=\frac{2}{3}\left(\frac{8\mathrm{T}}{\sigma}\right)^{1/2}\frac{D^{5/2}N_{\mathrm{H_2}}^4N_{\mathrm{CH_4}}^2}{\pi\varsigma^{3/2}c^{1/2}}\left(\frac{h^2}{2\pi m_{\mathrm{H_2}}\mathrm{T}}\right)^{15/4}\left(\frac{m_{\mathrm{CH_3}}}{m_{\mathrm{CH_4}}}\right)^3\upsilon'^2\upsilon^{3/2} \tag{38}$$

$$f_p=\left(2E_1+\Delta G_c+(3/2)q_1\right)+p\left(E_1-E_{\mathrm{btn}}^{(\mathrm{diss})}\right)/\mathrm{T}+\varsigma \tag{39}$$

$$\Delta G_c=(\sigma_1\varpi/4)(x_c)^4,\ x_c=(g_c)^{1/3}=\left(\sigma_1-3\varsigma\mathrm{T}+\sqrt{\sigma_1^2+3\varsigma\mathrm{T}\sigma_1}\right)/3\varpi\varsigma\mathrm{T} \tag{40}$$

are the values obtained in Section II, C_2^p are factorial multipliers in Newton binomial theorem, similar with (22), although the sum of exponents includes three members only.

The essential part of the integral of (37) at the supersaturation maximum moment is

$$\int_0^{t_\mathrm{m}}\exp\left(-\gamma(t_\mathrm{m}-\tau)^2\right)(t_\mathrm{m}-\tau)^{3/2}d\tau\approx\frac{\Gamma_{5/4}}{2\gamma^{5/4}} \tag{41}$$

$$f_p^{(\mathrm{m})} = f_p\left(\varsigma = \varsigma_{\mathrm{m}}\right),\ \gamma = (\alpha/2)\left(g_{\mathrm{c}}^{(\mathrm{m})}/2 - 1\right),\quad \alpha = \left|\varsigma''_{tt}\right|\left(t = t_{\mathrm{m}}\right) \tag{42}$$

are the values of the functions f_p at the maximum supersaturation moment t_{m}. The critical nucleation barrier and the critical nuclear size at this moment are:

$$\Delta G_c^{(\mathrm{m})} = \frac{1}{4}\sigma_1\varpi\left(x_c^{(\mathrm{m})}\right)^4,\ x_c^{(\mathrm{m})} = \left(g_c^{(\mathrm{m})}\right)^{1/3} = \frac{\sigma_1 - 3\varsigma_{\mathrm{m}}\mathrm{T}^{(\mathrm{m})} + \sqrt{\sigma_1^2 + 3\varsigma_{\mathrm{m}}\mathrm{T}^{(\mathrm{m})}\sigma_1}}{3\varpi\varsigma_{\mathrm{m}}\mathrm{T}^{(\mathrm{m})}} \tag{43}$$

Like the problem considered in Section II, the value of ΔN_1 at the moment t_{m} does not matter, because this moment itself is set by "hands." The supersaturation decay rate at the moment when $\Delta N_1 = N_1^{(\mathrm{source})}(1 - 1/e)$ (i.e., the supersaturation S decays approximately "e" times) is calculated in the course of the decision. Herewith, the expression (41) should be multiplied by 2, as the integral is taken not from $(-\infty, t_{\mathrm{m}})$, but from $(-\infty, \infty)$.

The result of calculation by (19) differs from the less cumbersome expression, obtained for the Knudsen cluster growth regime in Section II, by numerical factor and exponent factor. Therefore, the change in the nucleation temperatures $\mathrm{T}_{\mathrm{cross}}$, as compared with the values in Section II, is up to tens of K (Figure 23). At $\dot{\mathrm{T}} = 1000\,\mathrm{K/c}$ and the activation energy of the final steps of desorption $q_{\mathrm{a}}^{(\mathrm{des})} = 6.0$ эВ the value of $\mathrm{T}_{\mathrm{cross}} = 1820\ \mathrm{K}$. This value exceeds the temperature of "survival" of the diamond clusters, estimated above.

However, the estimated $\mathrm{T}_{\mathrm{cross}}$ is almost directly proportional to $q_{\mathrm{a}}^{(\mathrm{des})}$, supposed in the estimates to be equal to the bottleneck energy $q_{\mathrm{btn}}^{(\mathrm{des})}$ of the small species desorption from the diamond cluster. In turn, $q_{\mathrm{a}}^{(\mathrm{des})}$ and $q_{\mathrm{btn}}^{(\mathrm{des})}$ depend on the chosen quantum chemistry method (semi empirical or DFT). If $q_{\mathrm{btn}}^{(\mathrm{des})}$ is decreased to 5.0 eV, the nucleation temperature decreases in the same proportion to ~1400 K. Even more, if $q_a^{(des)}$ is noticeably lower than $q_{btn}^{(des)}$ (Figure 23), the nucleation temperature can be as low as 1000 K.

What is the most substantial here, it that the “graphite branch” of the dependence $\left|(d\varsigma / d\mathrm{T})_{expl}\right|$ is not realized at all and there is no need to analyse the competition between the onset of the graphite- and diamond nucleation. This allows the value of the temperature rise rate $\dot{\mathrm{T}}$ to be decreased. Therefore, the temperature of 1400 K, defined above as the upper limit, is quite real and may be the purpose of experiments.

Of course, the presented idea is far from self-consistent picture of the processes in the carbonate - silicate melt layer as an environment for the nucleation and growth of diamond clusters. For example, the number of nuclear formed was not calculated. Meanwhile, this value determines the Ostwald ripening after the completion of the explosive nucleation stage. Ideally, the process should be organized in such a way that a small number of big diamond clusters have survived and grew after dissolution of most of small clusters.

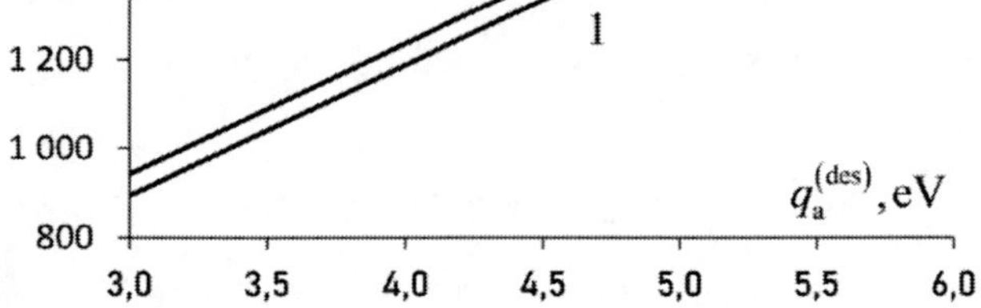

Figure 23. Diamond nucleation temperature depending on the activation energy of the final steps of the desorption process $q_a^{(des)}$, considered as a free parameter, at two very different temperature growth rates: 100 K/s (1) and 1000 K/s (2).

We did not try to optimize yet the pressure and the gas composition of the original gas mixture. Perhaps it is this factor that determines the number of nucleated and preserved diamond clusters.

CONCLUSION

In the first two Sections, the problem was put to trace the mechanism underlying the criterion of rapidly increasing temperature, which can be observed in the known diamond growing techniques. Initially, we tried to understand the sense of this criterion within the diamond CVD-growth technique stimulated in one way or another (MWCVD, HFCVD).

The analysis was started with thermodynamic properties of small carbon (diamond- and graphite) clusters and elementary processes of the gas mixture evolution leading to the formation of such clusters. It is shown, in particular, that the diamond clusters with hydrogen atoms adsorbed upon them are less beneficial than the graphite clusters of the same size.

The main channel of destruction of both types of carbon clusters is desorption of methyl CH_3-, ethyl C_2H_4 radicals and of the other small hydrocarbon species from the surface of clusters. At the same time, the bottleneck energy of desorption reaction sequence from the surface of the diamond cluster is only slightly more than the similar values for the graphite cluster surface. Much greater is the difference in the oxidative desorption kinetics.

The activation energy required to generate the clusters from the initial gas mixture in the carbon clusters nucleation reaction is about the same as the desorption energy for the same radicals as desorbed from the clusters (or only slightly exceeds it). Therefore, with a very slow temperature rise, weak supersaturation and carbon clusters nucleation are possible, but the process should be very weak as well, and the graphite must be synthesized first of all.

On the contrary, with rapid increase in temperature, the multi-stage desorption of the small species from the carbon cluster surface is significantly delayed as related to the temperature rise and can form a significant dynamic supersaturation that can lead to the diamond nucleation.

The solution to the problem of the crystal carbon cluster (diamond and graphite) nucleation from the gas mixture with rapid temperature rise showed that the nucleation really occurs. This means that there is a time moment when supersaturation of the gas mixture with the nucleation building material, which decays sharply due to the rapid temperature rise, decreases slower, but still remains huge by absolute value.

As the differences in the desorption dynamics of small hydrocarbon species from the surface of the diamond- and of the graphite clusters is negligible, nucleation of these types of clusters is almost simultaneous. At the same time, noticeable difference in oxidative desorption energies between them leads to a significantly higher level of the diamond cluster survival.

The substrate is not involved the diamond nucleation process in the initial consideration of the MWCVD method. The diamond nuclei that arise in the gas phase are deposited on the substrate as prepared clusters. However, the fast-heating criterion plays an important role in the diamond CVD-synthesis on any substrates. Therefore, considering the ideology of the CVD-synthesis of massive diamonds of high quality fundamentally exhausted, this criterion can be applied this criterion to the other possibilities of producing diamonds. The idea of non-stationary temperature dynamics and possibility of using it for gas-phase condensation control can work in conditions of such synthesis not on a substrate, but in the bulk of the medium that excludes the free growth of the diamond.

Our analysis showed that the carbon cluster surrounded by carbonate-silicate melt and arising from a gas-phase nucleation source, tends to develop as the diamond-, and not as the graphite cluster.

Experimental conditions under which the fast heating criterion “works” in the carbonate-silicate melt layer with the gas mixture passed through it, can be implemented.

The most significant of these conditions - the temperature and the required rate of the temperature rise from the initial temperature corresponding to the melt to the nucleation temperature of diamond clusters - are achievable under conditions of non-stationary HF induction heating.

The nucleation temperature is slightly different from that found in Section 2 for the MWCVD process in the same simplifications. This temperature is about 1800 K; however, this estimate is based on a number of parameters obtained from quantum-chemical calculations, and can actually by significantly smaller (about 1400 K). In any case, the temperature window of the cluster nucleation and at the same time their survival with respect to gasification is quite narrow and must be determined experimentally.

SUPPLEMENTS

1. The sense of intermediate steps in the reaction sequence steps shown in Figure 7.

The reactions shown in Figure 7, for the desorption process from the diamond cluster have the following topo-chemical sense:

1. $H_2 \leftrightarrow 2H$ (hydrogen molecule dissociation);
2. $C_nH_m + 2H \leftrightarrow C_nH_{m+2}$ (C–C- bond breaking and occupation of the broken valences;
3. $H_2 \leftrightarrow 2H$ (hydrogen molecule dissociation);
4. $C_nH_{m+2} + H \leftrightarrow C_{n-1}H_m + CH_3$ (desorption of the methyl radical and occupation of the broken valence on the atom 1);
5. $H_2 \leftrightarrow 2H$ (hydrogen molecule dissociation);
6. $C_{n-1}H_m + 2H \leftrightarrow C_{n-1}H_{m+2}$ (occupation of the broken valence on the atom 1);
7. $C_{n-1}H_{m+2} + 2H \leftrightarrow C_{n-2}H_m + CH_3$ (desorption of the methyl radical);

2. The sense of the first intermediate steps in the reaction sequence shown in Figure 12:

1. $C_nH_m + H_2O \leftrightarrow C_nH_{m-1}OH_2 + H$ (attachment of water molecule);
2. $C_nH_{m-1}OH_2 + H \leftrightarrow C_nH_mOH_2$ (C-C-bond breaking and attachment of H-atom to the free valence);
3. $C_nH_mOH_2 \leftrightarrow C_{n-1}H_m + COH_2$ (desorption of oxidized carbon within the methanal molecule);
4. $COH_2 \leftrightarrow CO + H_2$ (dissociation of methanal);

3. The rate of multistage reaction of the small species desorption from the cluster surface.

The equation for relative (with respect to equilibrium) concentration y_s of an intermediate reaction product at the stage "s" can be written in the diffusion approach as

$$\frac{\partial}{\partial t}\left(N_s^{(0)}y\right) = \frac{\partial}{\partial s}\left(w_s N_s^{(0)}\frac{\partial y}{\partial s}\right) \tag{1s}$$

where w_s is the direct reaction rate at the stage "s," $N_s^{(0)}$ – is the equilibrium cluster density at this stage. If it is taken into account that $\partial \ln N_s^{(0)}/\partial t = \left(q_s/T^2\right)\dot{T}$; $\partial \ln\left(w_s N_s^{(0)}\right)/\partial s = -\left(\partial\left(q_s + q_{a,s}\right)/\partial s\right)/T$ (where q_s is the total binding energy of the product at the stage "s" as related to the reactant "0") and the Zeldovich' solution is accepted for estimates [17, 34], i.e. the solution to (1s) without the left part, then the orders of magnitude of y and its derivatives in the region of the bottleneck s_{btn} is:

$$y\left(s_{btn}\right) = \left(\int_{s_{btn}}^{\infty} ds\left(wN_s^{(0)}\right)^{-1}\right)\left(\int_0^{\infty} ds\left(wN_s^{(0)}\right)^{-1}\right)^{-1} \sim \frac{1}{2},\ \partial y/\partial s \sim 1/2s_{btn},$$

$$\partial^2 y/\partial s^2 \sim 1/2s_{btn}^2.$$

Since s_{btn}>>1, the member with the first derivative in the right part is dominating: $\left(\partial \ln\left(w_s N_s^{(0)}\right)/\partial s\right)\left(\partial y/\partial s\right) >> \partial^2 y/\partial s^2$, and with $q_s > T$ (1s) turns into the first order equation on (*t*,s):

$$\dot{y} + \frac{q_s^{(des)}}{T^2}\dot{T}\,y = \frac{(d/ds)\left(q_s^{(des)} + q_a^{(des)}(s)\right)}{h}\exp\left(-\frac{q_a^{(des)}(s)}{T}\right)\frac{\partial y}{\partial s} \tag{2s}$$

where the symbols *q* additionally equipped for certainty by upper symbol "des," pointing at the desorption process. If it is accepted, as in (10), that the derivative $\dot{y}$ can be discarded in the left side of (2s), an ordinary differential equation on s is obtained. With $q_s^{(des)} > T$ its solution is

$$y(s) \approx y(s=0)\exp\left(-\frac{h\dot{T}}{T^2}\int_0^s \frac{q_{s'}^{(des)}}{(d/ds')\left(q_{s'}^{(des)} + q_{a,s'}^{(des)}\right)}\exp\left(\frac{q_{a,s'}^{(des)}}{T}\right)ds'\right) \tag{3s}$$

in the bottleneck region it should be set s=btn. The condition for the derivative $\dot{y}$ to be neglected is determined by the condition

$$\dot{y}/\left(\left(q_s^{(des)}/T^2\right)\dot{T}y\right) \approx \left(h\dot{T}q_a^{(des)}s_{btn}^2/2q_s^{(des)}T^2\right)\exp\left(q_a^{(des)}/T\right) << 1 \tag{4s}$$

With activation energy of 5 eV, this condition is valid at temperatures above 0.15 eV. This temperature lies in the area of nucleation, which is of interest for the problem solved. Providing that the condition, opposite to (4s), is true, the solution has the form of arbitrary function of the argument

$$s - \frac{\left(dq_s^{(des)}/ds\right)T^2}{h\dot{T}q_{a,s}^{(des)}}\exp\left(-\frac{q_a^{(des)}}{T}\right) \tag{5s}$$

e.g., $y(\phi) = \exp(-\phi^2)$, which decays at the initial moment by e^2 times at the interval $0 < \phi < 1$. The dependence of this form is easy to be sewn with (3s), e.g., at the interval of $\left(h\dot{T}q_a^{(des)}s_{btn}^2 / 2q_s^{(des)}T^2\right)\exp\left(q_a^{(des)} / T\right)$ from ½ up to 2 by the value and its derivative.

Competition Financial Interests

The authors have no competition financial interests to declare.

REFERENCES

[1] Shikata S. 2016. "Single crystal diamond wafers for high power electronics." *Diamond & Related Materials* 65:168–175.

[2] Umezawa H. 2018. "Recent advances in diamond power semiconductor devices." *Materials Science in Semiconductor Processing* 78:147–156.

[3] Alekseyev N. I., and Luchinin V. V. 2020. Electronics of Diamond. St. Petersburg: ElectroTechnical Univ.

[4] Koizumi S., H. Umezawa. J. Pernot, and M. Suzuki. 2019. "Power Electronics Device Applications of Diamond Semiconductors." *Woodhead Publishing Series in Electronic and Optical Materials* 468.

[5] Kukushkin S. A., and Osipov A. V. 2019. "Microscopic Description of the Mechanism of Transition between the 2H and 4H Polytypes of Silicon Carbide." *Physics of the Solid State* 3:288–291.

[6] Kukushkin S. A., and Osipov A. V. 2012. "Thin film heteroepitaxy through dilatation assemble formation." *Academy of sciences reports* 444(3):266-269.

[7] Kukushkin S. A., and Osipov A. V. 2014. "First-order phase transition through an intermediate state." *Physics of the Solid State* 4:792 – 800.

[8] Rudenko A. P, Kulakova I. I., and Scvortcovs V. L. 1993. "The chemical synthesis of diamond. Aspects of the general theory." *Russian Chemical Reviews* 62(2):87-161.

[9] Bundy F. P., W. A. Bassett, and M. S. Weathers. 1996. "The pressure-temperature phase and transformation diagram for carbon." *Carbon* 34(2):141–153.

[10] Anthony T. R. 1990. "Metastable synthesis of diamond." *Vacuum* 41(4–6):356–1359.

[11] Bachmann P. K. 1994. Properties and Growth of Diamond. Edited by Davies Gordon. London: Institution of Electrical Engineers.

[12] Digonsky S. V., and Ten V. V. 2006. *Unknown Hydrogen, role of hydrogen in polymorphism of solids, solid-phase reduction of oxides and sintering of powders*. St. Petersburg: Nauka (Science).

[13] Garanin V. K., Digonsky S. V., and Kudrjavtceva G. P. 2006. "Natural Dianond Formation Model in Aspect of its Synthesis. Diamond formation hypothesis in meteorites, metamorphic rocks and kimberlites." *Izvestia Vusov (Proceedings of universities). Geology and Geological exploration* 2:8-14.

[14] Portnov A. M. 1979. "Fluid Diapiration as the Cause of Formation of Kimberlite Tubes and Carbonatite Arrays." Rep. to USSR Sci. Acad. 246(2):416–420.

[15] Hassouni K., Lombardi G., and Duten X. 2006. "Overview of the different aspects in modeling moderate H_2 and H_2/CH_4 microwave discharges." Plasma Sources Science and Technology 15:117–125.

[16] Goodwin D. G., and Butler J. E. 1998. "Theory of diamond chemical vapour deposition." Edited by Prelas M. A., Popovici G., Bigelow L. K. Handbook of Industrial Diamonds and Diamond Films 527–581.

[17] Zeldovich J. B. 1942. "To Theory of a New Phase Formation." *JETP* 12:525-530.

[18] Kukushkin S. A., and Osipov A. V. 1996. "Thermodynamics and kinetics of phase transitions of the first kind on the surfaces of solids." Khimicheskaya Fizika (*Chemical physics*) 15(9):5-104.

[19] Chemistry Software Ltd. 1999. "HyberChem." Accessed 2017. http://www.chem.ac.ru/Chemistry/Soft/hyperchm.ru.html.

[20] Kukushkin S. A., and Osipov A. V. 1998. "Thin film condensation processes." *Uspehi Fizicheskih Nauk* (*Advances in physical sciences*) 168(10):1083-1116.

[21] Raizer J. P. 1959. *JETP*. 37:6(12):1741-1750.

[22] Zhukhovitskii D. I., Khrapak A. G., and Yakubov I. T. 1983. "Slag condensation kinetics in the channel of an mhd generator – explosive condensation." *High Temperature* 21(5):758–764.

[23] Zhukhovitskii D. I., Khrapak A. G., and Yakubov I. T. 1983. "Slag condensation in a MHD generator channel – quasistationary condensation." *High Temperature* 21(6):920–927.

[24] Johnson J. C., and Hinnov E. 1973. "Ionization, recombination and population of excited levels in hydrogen plasmas." *JQSRT* 13:333.

[25] Butler J. E., Mankelevich Y. A., and Cheeseman A. 2009. "Understanding the chemical vapour deposition of diamond: recent progress." *Phys. Condens. Matter* 21:1–20.

[26] Hassouni K., Lombardi G., Duten X., Haagelar G., Silva F., Gicquel A, Grotjohn T. A., Capitelli M., and Ropcke J. 2006. "Overview of the different aspects in modelling moderate pressure H2 and H2/CH4 microwave discharges." *Plasma Sources Sci. Technol.* 15:117–125.

[27] Zhimulev E., Shein M., Pokhilenko N. 2013. "Diamond crystallization in the Fe-S-C system." Nauki o Zemle (*Earth Sciences*) 451(1):729-731.

[28] Yamaoka S., Shaji Kumar M. D., Akaishi M., and Kanda H. 2000. "Reactions between carbon and water under diamond-stable high pressure and high temperature conditions." *Diam. Relat. Mater.* 9:1480 – 1486.

[29] Litvin Yu. A., and Zharikov V. A. 2000. "Experimental modeling of diamond genesis: Diamond crystallization in multicomponent carbonate-silicate melts at 5-7 GPa and 1200-1570°C." Earth Sciences 373(5):867-870.

[30] Palyanov Y., Sokol A. G., Borzdov Y. M., and Khokhryakov A. 2002. "Diamond formation through carbonate-silicate interaction." American Mineralogist 87(7):1009-1013.

[31] Sokol A. G., Palyanov Y., and Pal'yanova G. 2001. "Diamond and graphite crystallization from C–O–H fluids under high pressure and high temperature conditions." Diamond and Related Materials 10(12):2131–2136.

[32] Lebedev A. S., Eremyashev V. E, and Trofimov E. A. 2019. "Thermodynamic Analysis of the Interaction of Components in the Si–C–O System in the Carbothermic Synthesis of Silicon Carbide." *Doklady Chimii* (*Chemical reports*) 484(2):41-43.

[33] Rudnev V., 2003. *Handbook of Induction Heating.* New York: Marcel Dekker. 797.

[34] Lifshitz E. M., and L. P. Pitaevskii. 1980. *Statistical Physics, Part 2: Theory of the Condensed State.* Oxford, Butterworth-Heinemann.

In: Chemical Vapor Deposition (CVD) ISBN: 978-1-53619-949-9
Editor: Levi Karlsson

Chapter 3

GRAPHENE PREPARATION BY CHEMICAL VAPOUR DEPOSITION METHOD AND ITS BASIC PARAMETERS

P. Machac*
Department of Solid State Engineering,
University of Chemistry and Technology, Prague, Czech Republic

ABSTRACT

The presented text describes the properties of graphene, its preparation by a number of methods with the focus on the classical Chemical Vapour Deposition Method (CVD) on copper foil together with graphene transfer onto a dielectric substrate. It presents the results published by the author in a number of cited publications. It is thus an overview work, which can serve as a study material for those, who are beginning to get acquainted with graphene, its properties and preparation by the CVD method, which is frequently used in practice.

---

* Corresponding Author's Email: petr.machac@vscht.cz.

Keywords: graphene, chemical vapour deposition method, copper foil, dielectric substrate

1. Introduction

This chapter deals with graphene, its properties, and preparation with the focus on the Chemical Vapour Deposition (CVD) method and modifications thereof. The CVD method is one of the most important methods for graphene preparation; it is used in industrial preparation of graphene films. The chapter is amended with some partial results by the author in the given issue, which were published in further cited works.

The following abbreviations for various graphene forms will be used in further text for simplicity:

GR: general expression for graphene
MLG: single-layer graphene, which consists of just one monatomic layer of carbon
BLG: bilayer graphene
FLG: graphene consisting of several monatomic carbon layers; up to 10 carbon layers are still considered to be graphene in practice.

GR is a two-dimensional (2D) structure prepared from carbon atoms bound in crystalline lattice as per Figure 1. This is a relatively new material, which has been theoretically known and studied for about 60 years. In the beginnings that was basic research without applications, since the first GR samples were of microscopic dimensions and seemingly practically inapplicable [1]. The situation has changed at the beginning of this millennium, when the first usable samples were prepared; applications did not take long to come [2]. A patent was filled in 2003 for GR application is electronics [3]. In 2010, the development in GR research was crowned with the award of Nobel Prize in Physics to Russian researchers A.K. Geim and K.S. Novoselov, who succeeded in GR preparation by a simple method of mechanical exfoliation [3]. To the beginning of 2020 the SciFinder database records almost 300 000 references related to GR. Such a rapid growth of interest in GR is caused mainly by its exceptional properties, which make it useful for numerous applications of this material.

Figure 1. Idealised structure of MLG.

2. Properties of GR

In ideal case, GR consists of monatomic layer of carbon crystallizing in hexagonal system – see Figure 1 – where the distance of each atom from 3 neighbouring atoms is 1.42 Å [5]. The neighbouring atoms are bound with σ-bonds, which are formed by overlapping of hybridized orbitals derived from atomic orbitals s, p_x and p_y. The fourth atomic orbital p_z is oriented perpendicular to the GR plane. By hybridization p_z-orbitals the π-band and the π^*-band are formed, which lay over or under the crystal plane, respectively. These bands are a prerequisite for the exceptional electronic properties of GR.

The band structure of GR was first calculated in 1947 [6]. The conductivity and valence bands touch at six Dirac points. GR is thus a semiconductor with zero width of the band gap. In the points of contact, the dependence of energy on momentum is linear, and in case of fault-free GR the effective weight of charge carriers is practically zero. Thanks to this, the mobility of charge carriers is very high and may reach up to 15 000 cm^2/Vs at room temperature. Some cited literature reports the value up to 200 000 cm^2/Vs [7]. Thus, movement of charge carriers in GR occurs by ballistic transport.

Other excellent properties of GR include their thermal conductivity, which may reach up to 3000 – 5000 W/mK [8]. If, however, GR is fixed to the substrate, then this value will be lower. From electronic properties, result important optical ones, which make GR an important material for optoelectronic applications. It is above all high transparency, which reaches the value approx. 93.7% for the visible radiation for MLG [9]. This value predestines GR to be used as a transparent conductive electrode. To other significant properties of GR belongs high mechanical tensile strength, which reaches the value of 130 GPa and Young module of 1 TPa [10].

Chemical properties of graphene are highly important, too. GR as such consists of atoms of carbon, which is chemically stable by itself and thus it is little reactive. GR has, however, very large specific surface of approx. 2600 m^2/g [11], which predisposes it for construction of sensors. For these purposes, a GR film must have suitably bonded functional groups, by either chemical or physical adsorption. As a result, the electronic properties will change. Whereas every GR atom is a surface one at the same time, the prepared structure will show high sensitivity to bound analyte.

Due to the number of above-listed properties, GR has found possibilities for applications in many areas of technology [12]. High mobility of charge carriers predisposes GR to construction of field effect transistors (FET). A certain problem is posed by the fact that GR is of zero width of the band gap and transistors utilizing such semiconductive material for construction of the channel shall have low ratio of channel resistance in the off-state to that in the on-state. This fact is unacceptable for application of GR transistors in digital technology. The situation can be solved by constructing the transistor channel of so-called nanoribbons (narrow strips of GR 1-2 nm wide), which may have a band gap up to 1 eV [13]. A certain possibility is also the use of BLG, which may show a band gap under certain conditions [12].

In a number of optoelectronic structures, it is necessary to have light-transmitting conductive electrodes available (electroluminescent diodes, solar cells, displays, screens, etc.). Historically, a mixture of tin and indium oxides (ITO) as well as some other materials [9] has been used here. GR seems to be a replacement material, which has very low absorption of visible light, very low electric resistance and good mechanical properties (strength and flexibility).

GR has found an application in the area of energy storage, too – being used in lithium-ion batteries and supercapacitors. The first application utilizes a layer of modified GR, which serves as anode of a cell and enables reaching higher capacity of the battery [14]. In the second application, large area of a GR layer of which one of the capacitor electrodes is made, is used for increasing the capacity of the capacitor [15].

GR with its chemical properties began to be used in sensor technology as well [12]. GR layers may serve as a base of detectors of various gases, as well as other special substances, such as various biomaterials and similar. The field of application of such sensors is above all in medicine.

There are certainly more areas, where GR is already in use or where its use is assumed, than shown in this short survey. They include for example protection of materials from corrosion, preparation of various composite materials, spintronic, and many others. For potential applications of GR will be in largely decisive preparation of GR in industrial scale, its properties and, last but not least, the production costs, which will be ultimately the decisive factor for massive deployment of this promising material.

3. Methods of GR Preparation

In this chapter, significant methods of GR preparation that are used in laboratory and eventually also in industrial practice will be briefly discussed.

The methods of GR preparation are in principle divided into two classes: Top-down and Bottom-up. In case of the first type, it is the growth of GR from pre-prepared material containing carbon, such as graphite or graphite oxide [12]. Among the Top-down methods belongs namely graphene growth by mechanical exfoliation. Based on its use, Geim and Novosedlov were awarded with Nobel Prize [3]. The principle of this method is very simple; it consists in gradual repeated reduction of number of monolayers in a graphite crystal by means of Scotch tape. Optically transparent flakes, which get caught on the Scotch, are released from the tape by dissolving the tape in acetone, and sediment on a silicone substrate. This method does not require any complicated equipment; it provides high-quality GR, but only in size of several micrometers. From this method arises the method of graphite sonication [12]. Here, graphite powder is used, which is inserted into sonication bath of various chemicals. After long-term sonication, depending on the type of chemical used, it is possible to obtain small GR flakes. Top-down methods further include the process of graphite oxide reduction [12]. Reduction of graphite oxide is carried out chemically i.e., in hydrazine or thermally, by annealing in argon/helium atmosphere [16]. In both cases, the result is reduced GR oxide, which has worse parameters than mechanically exfoliated GR. This material, however, suffices by its parameters for partial applications. This group of methods further includes the method of carbon nanotubes opening [17]. This method, too, utilizes the principles of graphite sonication.

Methods of GR preparation by the Top-down process belong to laboratory processes, which cannot provide MLG in large areas. For a number of applications, however, GR layers of large surface consisting of large monocrystallic areas are necessary. In contrary, the Bottom-up methods are possible to use for preparation of MLG by much more reproductive ways in relatively large areas. As such, these methods are usable for GR preparation for practical applications. In the Bottom-up methods, the carbon films grow in cooperation with catalytic layer by gradual deposition of carbon atoms on the surface of a catalyte, or possibly by carbon diffusion, its sublimation or segregation. Some of these methods are predestined for growth of MLG or formation of FLG [12]. One of the basic Bottom-up methods is the already mentioned CVD method and its modifications [18], to which the following chapters will be dedicated.

A significant method of GR preparation is epitaxy growth on the surface of SiC [12]. This method is based on thermal decomposition of SiC at high temperature, when silicon as the more volatile constituent sublimes and the remaining atoms of carbon form at suitable conditions a GR layer on the surface. A drawback of this method is the high temperature required for GR formation (up to 1500°C) and expensive substrate material. If we however use the semi-insulating SiC, we obtain directly the GR layer on dielectric substrate. A modification of this method can be GR growth by so-called silicidation of metallic layer on SiC at relatively low temperature [19]. This method is primarily based on the Ni/SiC structure, by annealing of which at temperatures 800 – 900°C reaction of metal and SiC takes place at formation of silicides and free atoms of carbon at the metal/SiC boundary. The atoms of carbon diffuse through the layer of metal up to its upper surface. At the subsequent rapid cooling of the structure, carbon segregates due to substantially lower solubility of carbon in metal on its surface as well as on the boundary of metal and SiC, where thin carbon layers are formed. These layers correspond usually by its parameters to FLG. A similar method of GR growth is based on the metal/carbon/SiO_2 structure [20]. The metal can be again nickel or cobalt. Carbon then enters the structure in the form of thin layer of pure carbon or in the form of polymer. At annealing of the structure diffusion of carbon atoms into the layer of metal takes place again, and at the subsequent rapid cooling carbon atoms segregate from the metal out, similarly to the previous method. After etching-off the metallization we obtain the required GR/SiO_2 structure. Prepared GR is usually again of FLG. The advantage of this method is direct preparation of GR on dielectric substrate. Another modifications of the described method consist in use of metallic substrate – GR is here prepared from a carbon/metal structure [21]. The process is very simple, but if we need to form a structure of GR/dielectric material, then we have to supplement this method with suitable transfer of the GR film – the process is similar to that at the CVD growth and it will be described further in the respective chapter.

There are certainly more methods of GR preparation; the submitted survey lists the most significant ones, which are used both in laboratories of universities and research centres and in commercial growth of GR.

4. CVD Preparation of GR

The method of Chemical Vapour Deposition is currently one of the most widely used methods for production of GR layers on the surface of various metals [18]. As mentioned earlier, CVD belongs to the Bottom-up methods. The source of carbon is a gas (most frequently methane or ethyn), which decomposes at increased temperature on a surface of metal. Atoms of carbon get absorbed on the surface and form a layer of GR. The classical CVD method uses metals such as copper, which shows minimal solubility for carbon and limits formation to just one layer of GR. Other metals, such as nickel, cobalt, palladium, iridium or ruthenium were tested as well, however FLG gets formed on these metals.

Molecules of gas fall onto the surface of the catalytic metal; they get decomposed here at high temperature. Carbon atoms diffuse on the metal surface and consequently growth of GR crystals takes place on nucleation centres, where the individual atoms reach the boundary of emerging monocrystallic domain. The size of crystals of the resulting GR therefore depends on density of nucleation centres; the more nucleation centres are presented; the more crystals get formed. Nucleation is a random process; so the faster the growth of GR is, the lower is the density of the nucleation centres and the larger domains are formed. More nucleation centres are formed at slow growth; the prepared layer of GR will have more failures and domains of smaller area get formed. In the points of contact of the individual GR crystals, occurrence of failures (grain boundaries) takes place. Surfaces of the crystalline domains may have dimensions ranging from μm up to cm.

Quality of catalytic substrate (most frequently copper) is of high importance at GR growth by the CVD method. The experiments suggest that there is a certain compromise solution between the growth speed of GR and size of the domains, which can be solved by composition of the metallic substrate. For example, it is possible to use alloy Cu-Ni [22] instead of pure copper. Nickel increases the catalytic activity of the substrate and thus it increases speed of GR growth and ensures large domains at the same time. Another possible solution is the choice of monocrystalline Cu substrate with in-plane surface (111) [23]. In this case, carbon atoms have low surface diffusion and growth of high-quality MLG takes place.

The CVD growth of GR is further influenced by the flow rate of the gas, which serves as a source of carbon, most frequently methane. The flow rate influences the structural defects and usually it has no effect on the number of GR layers. The flow is kept on low levels in the nucleation area of growth in order to achieve low density of nucleation centres and it is increased at the growth phase in order to achieve higher rate of GR growth [24].

In order to obtain higher quality of GR, hydrogen and sometime even oxygen are being added to methane. Quantity of hydrogen influences the number of nucleation centres on the surface and thus the size of domains. At higher concentrations hydrogen causes etching-off of the already developed GR and decreases the quantity of randomly formed nucleation points [25]. Usually it is necessary to optimise the share of hydrogen in methane. Growth rate of GR is influenced even by a small quantity of oxygen in the gas mixture. Oxygen reduces the number of nucleation centres by passivation of copper surface and enables formation of GR with domains of size up to several cm^2; the quantity of oxygen, however, must be optimised as well, since higher share of oxygen causes growth of polycrystalline GR [26].

Last, but not least, the growth quality is affected by the growth temperature. High temperature is necessary for formation of high-quality graphene as it ensures sufficient dissociation of methane, diffusion of carbon atoms and GR growth rate. With increase of the growth temperature, density of the nucleation centres decreases; therefore, better GR growth takes place [27]. High temperature may, however, cause copper sublimation and local failures of the GR film [28].

The basic advantage of the CVD growth of GR is – as mentioned above – preparation of large-area MLG films. The method is now used as a manufacturing method for numerous GR applications in technical practice. The drawback of this method is, that the resulting GR layer is placed on a conductive substrate – most frequently on copper foil. For majority of applications it is however required, that the obtained GR layer lays on a dielectric substrate [12]. This is why various methods of so-called GR transfer onto a dielectric substrate are used. This is a significant technological problem, which must be solved by GR producers.

There is a certain alternative to the CVD growth of GR – so called plasma enhanced CVD (PECVD) growth [29]. At this method, decomposition of gas, which is bringing carbon into the reactor (such as methane), is carried out by means of plasma. The principal advantage of this method is the possibility of growth at far lower temperature than that of the classical CVD method. The PECVD growth of GR can be carried out at temperatures lower than 500°C. Another advantage is direct preparation on non-metallic substrates and therefore it is possible to combine this technology with other technologies i.e., in the manufacturing of CMOS elements. The basic disadvantage of the PECVD method is substantially higher failure rate of the prepared GR as well as the fact that plasma may cause damage to the substrate.

For preparation of GR by the classical CVD method it is necessary to use the GR transfer, as mentioned above. In practice, there have been developed a number of procedures for GR films from metallic foils onto a suitable dielectric pad. This process may have significant effect onto the parameters of the resulting product.

Methods of GR transfer can be divided to dry and wet ones [30]. Both processes begin with etching-off the copper foil, on which GR has been prepared. If GR after etching-off is in contact with the new substrate, then it is the dry transfer. In case that GR after etching-off swims in a liquid (i.e., in water) and a liquid is required for the transfer in order to get attached on the new substrate, then it is the wet transfer. Generally, we can say, that if more steps are required within the process where liquid is involved, then the resulting GR is less contaminated and less faulty [3]. The general scheme of the transfer consists in that an auxiliary layer of material is applied onto GR, copper foil gets etched-off, GR is attached onto the required substrate and the auxiliary layer is removed.

The process of the wet transfer is schematically illustrated at Figure 2 [32]. This method generally transfers the GR film onto polymer substrate PMMA reinforced with PDMS, with the whole structure being glued on SiO_2 (300 nm)/Si. Figure 2(a) shows the initial GR produced by CVD growth on copper, where GR is formed on both sides of the Cu foil. Subsequently, deposition of PMMA and PDMS layers follows, on one side of the structure – see Figure 2(b). The PDMS solution was prepared by mixing in ration 10:1 with the respective hardener and its drop (depending on the used area) was applied onto the PMMA surface [32]. Before the PDMS cured, substrate plate SiO_2 (300nm)/Si was pressed onto it – see Figure 2(c). After the PDMS layer cured (at 100°C for 90 min), it fused with the SiO_2/Si substrate – see Figure 2(d). Another essential step in the process was burning of the lower GR layer on Cu with oxygen plasma (pressure $1x10^{-1}$ mbar, power 60W, duration 12 min) for easier etching of copper – see Figure 2(e). The last step was etching-off of Cu foil in ammonium persulfate solution or in diluted $FeCl_3$, when the final structure of GR/PMMA/PDMS/SiO_2/Si – Figure 2(f) was obtained.

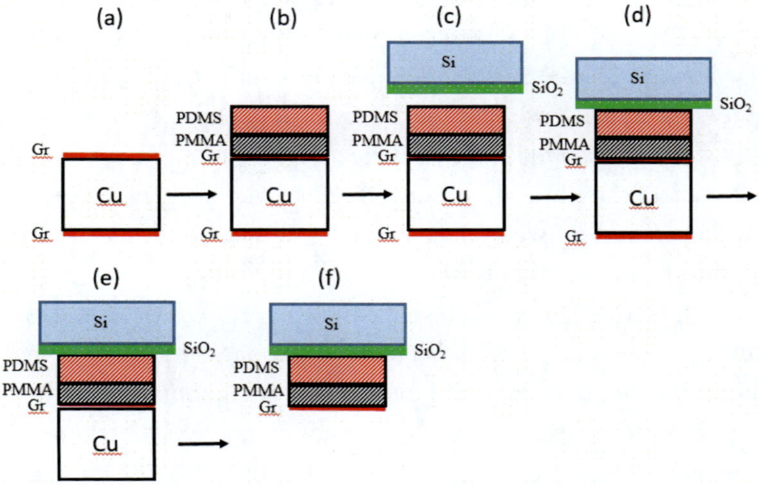

Figure 2. Wet transfer of GR, a – CVD growth of GR, b - Deposition of PMMA by spin coating, drying at 100°C, dropping of PDMS, c - Pressing of Si/SiO_2 on PDMS, d - Curing of PDMS at 100°C for 90 min, fusing with Si/SiO_2, e - Plasma burning of lower GR, f - Etching of Cu foil.

So-called thermal release tape [33] can be used for the GR transfer. This tape gets pressed against GR prepared by the CVD process on copper. The next step is etching-off of copper and then GR with the thermal release tape gets pressed against the carrying polymer, such as PET. The thermal release tape gets then released by heating and the resulting structure of GR on PET is obtained. This structure can be used for example as a basis for a touch display.

5. Transfer-Free CVD Process

As described in the previous chapter, the basic drawback of the classical CVD method of GR growth is the necessity of GR transfer onto a dielectric substrate. This transfer may deteriorate parameters of GR prepared by the CVD growth, since this is a technologically demanding operation – see the previous chapter. This is why methods of so-called transfer-free synthesis [33] were developed. The principal advantage of these methods is the fact that no residues of polymers or thermal release tape remain on the GR film.

The scheme of GR preparation by standard transfer-free CVD process [34] is shown at Figure 3. The principle consists in that a layer of catalytic metal (cobalt in this case, but nickel or nickel-copper combination can be used as well) gets applied onto a dielectric substrate (i.e., SiO_2/Si, silicon glass or other material), see Figure 3a. Then GR is prepared by the CVD process, whereas it is assumed that GR gets formed on the metal surface, but also that diffusion of carbon atoms through the metallic layer to the metal-dielectric material boundary takes place – see Figure 3b,c. During the cooling after finishing the CVD process, carbon segregates on the metal-dielectric material boundary due to decrease of its solubility in the metal – see Figure 3d. The next stage of GR formation process is etching off of the metallization, which can be preceded by the operation of burning-off the upper GR in oxygen plasma – See Figure 3e. Then we obtain the required structure of GR/dielectric material – see Figure 3f. As you can see, the process is not burdened by the necessity to transfer GR from the metal onto a dielectric substrate.

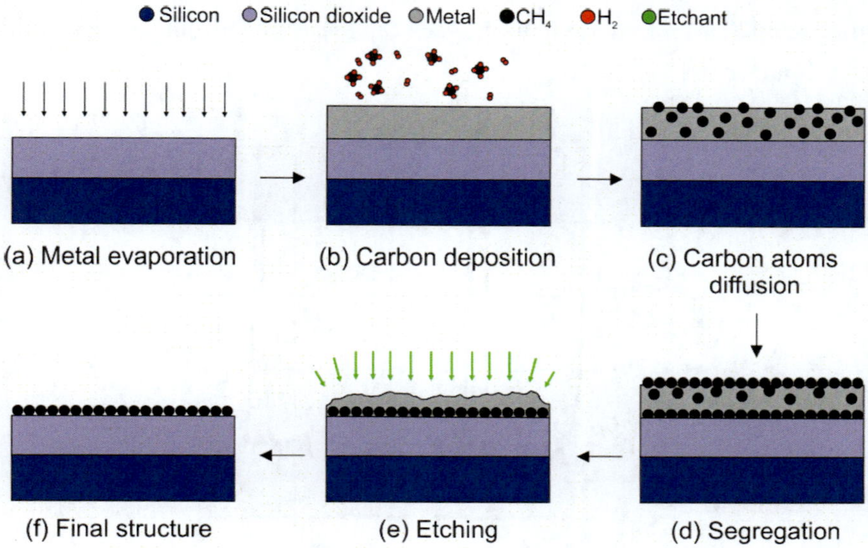

Figure 3. Scheme of the transfer-free CVD growth process.

The specific procedure of GR preparation as described in the work [34] is as follows:

- A layer of cobalt was deposited by an electron gun of thickness 50 nm.
- The CVD process took place in the nanoCVD apparatus (Moorfield Nanotechnology Ltd) with a cold-wall reactor for 480 s at the temperature of 900 °C, in mixture of gasses H_2:CH_4 in the ratio 2:35, at pressure of 1.33 kPa.
- The upper GR was burned-off by oxygen plasma (10 Pa, 30 min, 1.2 W/cm^2) and the subsequent etching of the metallic layer was carried out in citric acid.

Figure 4 shows an example of Raman spectrum of the obtained GR [34]. From the ratio of the 2D/G band it is possible to deduce that the GR consists of 2 monolayers of carbon [35], from the ratio of bands D/G can be deduced the size of GR crystals. Considering the variance of GR parameters on the substrate surface the size is 180 – 600 nm. Figure 5 then shows the GR surface on SEM micrograph, where the boundaries of individual micrograins of GR are seen [34].

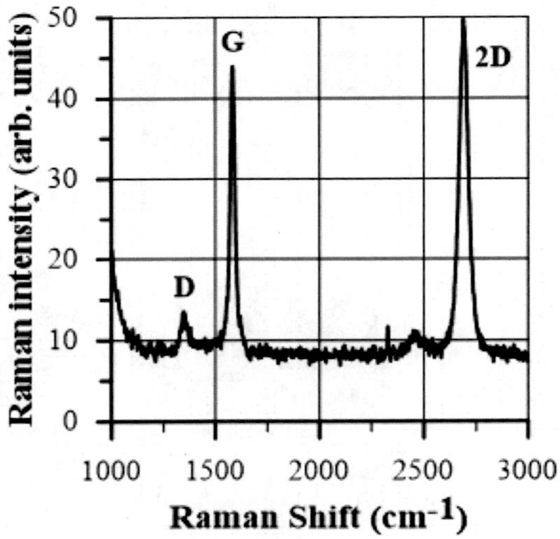

Figure 4. Raman spectrum of the best GR film.

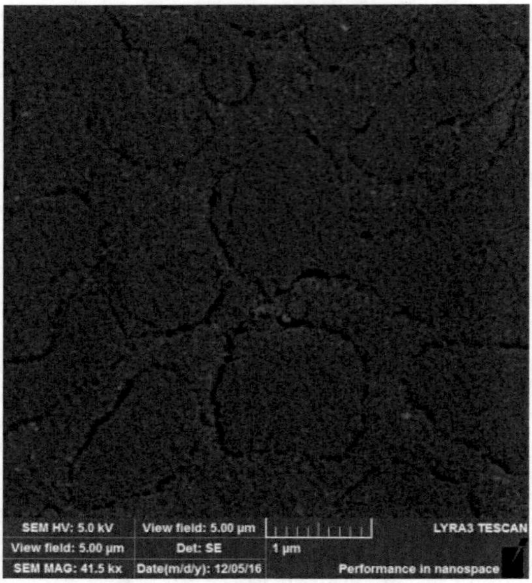

Figure 5. SEM micrograph of the GR film.

6. Surface Treatment of Cu Foil Before CVD Growth of GR

As mentioned already in the previous chapters, the CVD process belongs to the highly significant methods of GR growth. In chapter 4 it is described, that parameters of GR prepared by the CVD method depend on the surface of the catalytic metal (copper), on which the growth is done – a combination of copper and nickel, or the use of monocrystalline copper substrate respectively [22, 23]. However, a number of experimental works indicate that the quality of GR prepared by the CVD method can also be influenced by chemical treatment of the used copper foil. For example, experiments with electrochemical cleaning of Cu foil before CVD growth of GR by acetic acid were carried out; formation of high-quality MLG was reached [36]. Similarly, the surface of Cu foil was improved before CVD growth by electrochemical polishing in phosphoric acid at increased temperature [37].

Verification, whether the chemical treatment of Cu foil before CVD growth of GR improves the parameters of the prepared graphene, was carried out in our research establishment as well [38]. The work brings comparison of influence of various electrochemical processes for cleaning of Cu foils onto roughness of their surfaces and onto quality of GR prepared by the CVD process on these foils. The basic material for our experiments was copper foil by the AlfaAesar, 0.025 mm thick, Puratronic, 99.999% (metal basic). Altogether 13 electrochemical processes were tested. The basic chemical for all electrolytes was phosphoric acid with addition of other chemicals (acetic acid, polyethylene glycol, chromium oxide, ethanol and urea). The followed copper foil formed always anode in the electrochemical process; cathode was made of another Cu foil of roughly 4 times larger area. Direct current was supplied into the circuit by the current source in the range from 10 to 100 mA/cm^2. Quality of the foil surfaces was compared by measuring of roughness R_a using Laser Scanning Microscope. All experiments were carried out at room temperature (25°C). For comparison, roughness of unprocessed foil was taken as a standard. At Figure 6a there is the image of surface of unetched Cu foil (R_a = 1.13 μm) and at Figure 6b there is the image of surface of foil treated by the best electrochemical process No. 13 (30 ml H_3PO_4, 12 ml H_2O, 7 g CrO_3, 100 mA/cm^2, 1 min, R_a = 0.557 μm) [38]. It is obvious from the images that the process has smothered the foil surface.

GR growth was carried out in the nanoCVD-8G apparatus from MOORFIELD Nanotechnology. The growth conditions were following: temperature 1000°C, process time 10 min., flow ratio of gasses in the reactor H_2:CH_4 = 16:4 sccm, overall gas pressure 1.33 kPa. Transfer of GR from the Cu foils was done using polymethylmethacrylate (PMMA). Copper foils were etched off in aqueous solution of ammonium persulfate of 0.2 M concentration. GR was then transferred onto Si plates covered with a layer of SiO_2 300 nm thick and PMMA was subsequently removed by dissolving in acetone.

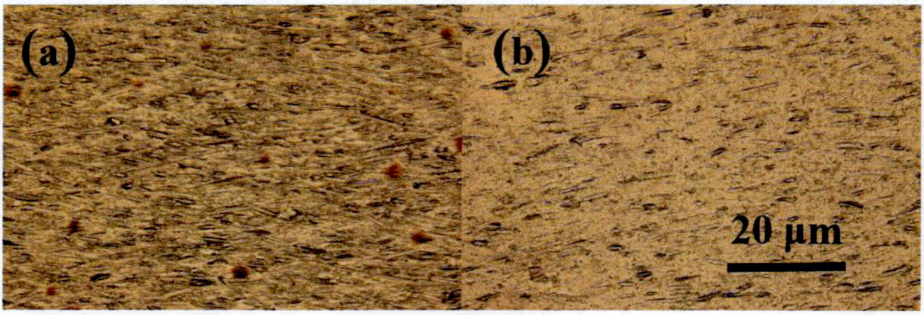

Figure 6. Surfaces of Cu foils: a – chemically untreated (reference sample), b – treated by the best electrochemical process No. 13.

Figure 7 shows Raman spectra of GR prepared on the reference sample, graphene on Cu foil treated by two chosen processes No. 1 (concentrate H_3PO_4, 30 mA/cm^2, 5 min) and No. 13 [38]. In all cases, it is GR transferred onto SiO_2. In case of GR on Cu foil treated by the process No. 13 the GR is monolayer (the ratio of peaks 2D/G reaches up to 2), GR prepared on the foil treated by the process No. 1 shows the lowest failure rate against referential graphene (I_D/I_G = 0.17). From the results, it is obvious that the electrochemical treating process influenced very strongly obtained results of GR prepared by the CVD growth method.

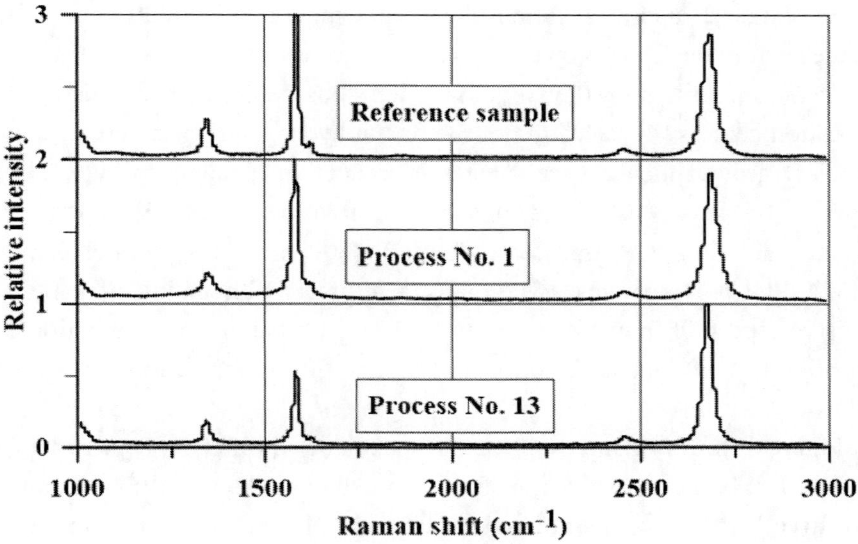

Figure 7. Chosen Raman spectra of GR films transferred onto SiO_2.

7. Parameters of GR Growth by CVD on Copper

This chapter presents results obtained by a number of analyses on GR prepared by the CVD process in our research establishment, where we have long term be engaged in GR preparation by various methods. As mentioned in the previous text, we have available the nanoCVD-8G apparatus (Moorfield Nanotechnology Ltd) with cold-wall reactor. The advantage of this apparatus is its relative simplicity and low operational costs, as the apparatus is equipped with heated table on which the growth takes place. The CVD reactor itself is cold, which saves a great amount of energy for heating at the growth (up to 90% savings against a classical CVD system) and it is possible to reach fast heating and consequently fast cooling [39]. The said CVD apparatus was used for preparation of GR on Cu foil, which was further transferred onto the SiO_2/Si substrate and subjected to measuring of a number of significant properties [40].

The process of GR preparation was supplemented by oxidation of Cu foil in standard laboratory atmosphere, at temperature of 200°C for 20 minutes before the CVD growth. This step positively influences the quantity of organic impurities on the Cu surface and thus decreases the density of nucleation centres [41]. Heating of the foil in hydrogen atmosphere during the CVD growth then causes substantial reduction of copper oxide. The growth of GR itself was carried out at conditions as described in the previous chapter, including GR transfer onto SiO_2 (300 nm)/Si substrate, using PMMA; the polymer was additionally reinforced with gluing of kapton adhesive tape 0.05 mm thick. Growth of GR further included additional annealing in UHV apparatus prior to analyses (500°C for 45 min, pressure bellow 5×10^{-6} Pa) [40].

Figure 8 shows examples of Raman spectra of GR after CVD growth on Cu foil (the black curve), GR transferred onto the SiO_2 (300 nm)/Si substrate (red curve) and GR on the same substrate after UHV annealing (blue curve). The spectrum of GR on copper is influenced by luminescence of the Cu foil; after the transfer of GR onto the SiO_2 (300 nm)/Si substrate peaks D, G and 2D are clearly visible and after the UHV annealing the crystallinity of the layer improved (ratio of the peak sized D/G decreased and that of peaks 2D/G increased).

Surface of GR was examined by means of Atomic Force Microscopy (AFM). Figure 9 shows the acquired images. Figure 9a shows details of the GR surface by AFM directly after the transfer to SiO_2. It is evident from the image that PMMA residues (small white points) remain on the surface after the transfer of GR onto SiO_2. Small imperfection of GR after the transfer can be seen on the surface too. A much better surface is shown on Figure 9b, which represents the image of the sample surface after the UHV annealing. The number of tiny formations and bodies in the form of PMMA remains and other contamination originating from the transfer process is greatly reduced. The morphology or structure of the GR layer itself in the as received state and after the UHV annealing appears very comparable.

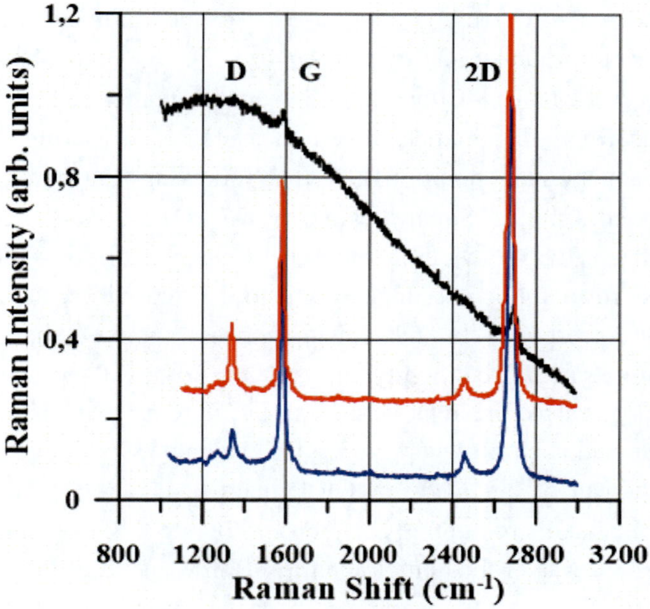

Figure 8. Chosen Raman spectra of GR films on Cu foil (black spectrum), on SiO_2 (300 nm)/Si substrate (red spectrum) and after UHV annealing (blue spectrum).

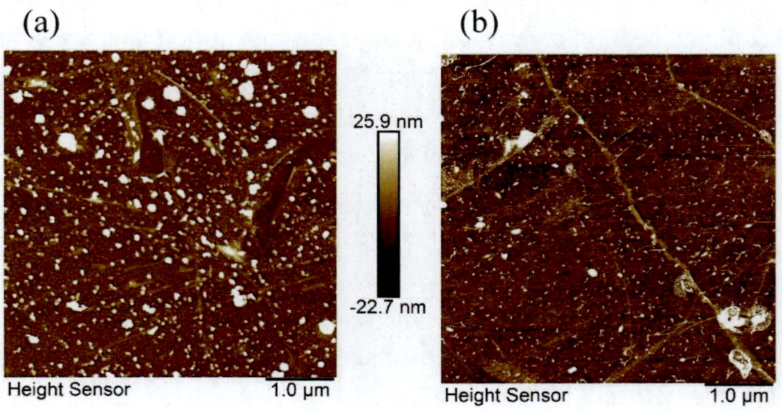

Figure 9. Surface of GR film visualised by AFM: prior UHV annealing (a), after UHV annealing (b).

Van der Pauw method was used for measurement of electrical parameters on an apparatus mentioned in [42]. Standard ohmic contacts Au(40 nm)/Cr(10 nm) [43] prepared by vacuum evaporation were used for these measuring. GR parameters were measured at measuring current of 5 μA and magnetic induction of 0.43 T. The following data were obtained on as transferred samples: Square resistivity: 57 kΩ, square concentration of charge carriers: 2.35×10^{11} cm^{-2}, mobility of charge carriers: - 2700 cm^2/Vs. From these results, it is possible to come to the following conclusion: high resistivity corresponds to low square concentration of charge carriers. Considering its negative mobility, the said GR is of low P-type conductivity. The mobility values are relatively high, but they are similar to published values obtained in similar works – see for example [39, 44].

The subject of our study of GR properties was further the effort to visualize the band structure of GR prepared by the CVD method [40]. The band structure was probed employing a k-PEEM mode of the NanoESCA instrument. We used the equipment of the cooperating establishment Institute of Physics of the Academy of Sciences, Czech Republic. The k-PEEM results of the sample after post annealing are presented in Figure 10. E vs. k cuts of the band structure were obtained and plotted along high symmetry points Γ, M and K, which were readily identified and which are typical for GR. At the K point, contours of the bottom part of the Dirac cone are obvious, reaching up to Fermi level. The Dirac point at Fermi level implies undoped GR layers. This result corresponds to the low doping of the GR film obtained from measuring of electrical properties as per the previous paragraph. The constant energy vs. k|| image from Figure 11 provides further information about the sample. Firstly, an apparent single set of K points without signs of replicas evidences a single orientation of GR layers of the measured region (a). However, a measurement in a different region on the sample reveals two sets of hexagons formed by K points (b). The mutual orientation of hexagons is from 25° to 30°. In fact, we should identify and we indeed do identify regions with MLG, BLG or folded GR layers originating from the transfer process. Further measurements on more regions of the sample surface bring the same results; only the intensity of individual hexagons varies.

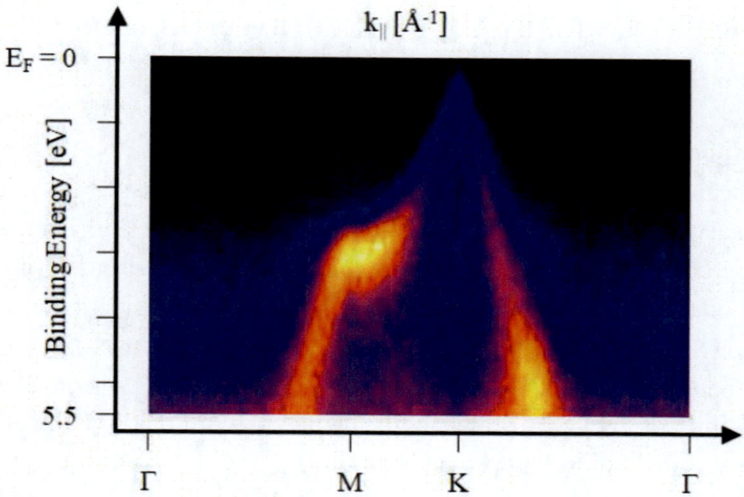

Figure 10. E vs. $k_{||}$ cut through band structure of GR layers along high symmetry points mostly used for presentation. Measurement was done after high temperature annealing.

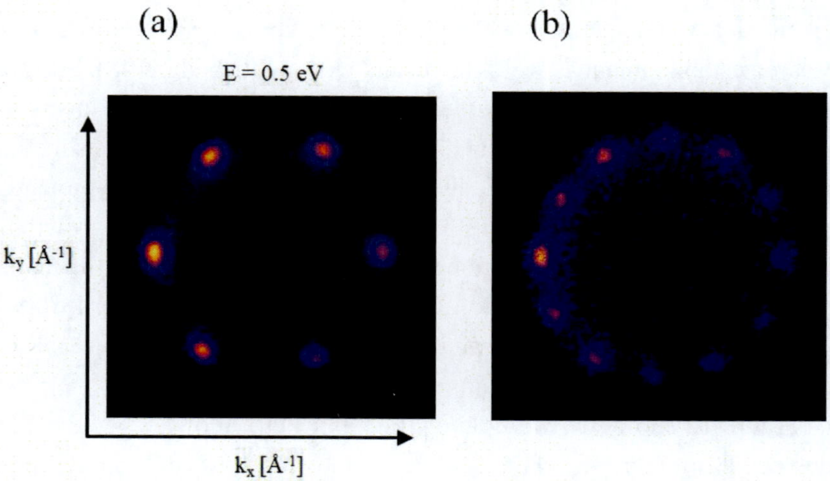

Figure 11. Constant energy cut of band structure of GR layers at 0.5 eV below E_F. A region with single rotation of graphene domains (a), a region with two rotations of GR domains (b). Lower signal intensity towards the bottom right region of the image is of instrumental origin.

8. GR Prepared by CVD Method on Silicon Carbide

This chapter describes the properties of GR prepared by the CVD growth on SiC substrate. The methods of GR preparation on SiC substrate have been mentioned in chapter 3. It is mainly the method of epitaxy growth (thermal decomposition of SiC at high temperature, when silicon as the more volatile component sublimes and the remaining carbon atoms form a layer of GR on the surface) [45]. Further, it is the method of silicidation of metallic layer on SiC at relatively low temperature (annealing of the Ni/SiC structure at temperatures of 800 – 900°C) [19]. The first of the said methods enables preparation of MLG, its drawback is the necessity of the high temperature. The other method usually provides FLG only.

However, literature searches show that it is possible to use also the CVD growth for preparation of GR on SiC substrate, using various gaseous sources of carbon and at various temperatures - toluene and 900°C [45], ethylene and 1200°C [46], propane and 1300°C [47]. With the above-described CVD apparatus we attempted to prepare GR on SiC in our research establishment as well; the results were published in [48]. Substrates of 4H SiC and 6H SiC type were used for the preparation. GR layers were prepared under the following conditions: the source of carbon was methane, or mixture of methane and hydrogen at various ratios, preparation temperature 1080°C, pressure 1.33 kPa, methane flow 18 sccm, growth time 300 – 4300s. Growth optimization was carried out by changing the methane to hydrogen ratio in the reaction chamber, by the CVD process time and finally by selection of SiC type and substrate face (Si face or C face). Raman and XPS spectroscopy was again used for the basic analysis.

Figure 12 shows an example of Raman spectrum of GR, which can be considered optimal [48]. This is a spectrum of GR prepared on 4H SiC, Si face growth time 3600s, and pure methane. From the spectrum it is evident that the prepared GR is highly faulty (significant amplitude of the D peak against the G peak) and FLG – peaks intensity ratio 2D/G = 0.47. This was a GR film of the best parameters; in numerous cases no 2D peak occurred at the Raman spectra at all, or just a small one. It was found that the Si face is more fitting for GR growth. Quality of prepared GR was just very slightly dependent on the type of SiC substrate.

For verification of the results of Raman spectroscopy, XPS spectroscopy was used [48]. From the obtained survey spectra – Figure 13, it was evident that only carbon with minimal quantity of oxygen (below one percent) was found on the surface of the samples. Very weak signal of silicon coming from the SiC substrate was visible in case of electron detection under perpendicular angle. Figure 14 shows detailed XPS spectrum of carbon C 1s. The black line represents the measured spectrum, the blue line represents spectrum obtained by fitting, when the asymmetric bonding of C=C and π=π was used (red curve - C=C bonds and green curve - π-π bonds). Violet curve represents background of the measurement. The band belonging to carbon from SiC substrate is not present within the spectrum, though there is very weak band corresponding to silicone within the survey spectrum. It is highly important, that the structure contains just a very small quantity of oxygen, which gives evidence that practically no oxidation of GR film takes place. The shape of Raman spectra of the formed GR films (Figure 12) may be nearing to the shape of GR oxide [49]. The XPS measurement, however, excludes this eventuality; therefore, it can be declared that carbon on the surface of SiC substrates is GR and not graphene oxide.

CVD growth on SiC substrate is generally similar to the method of epitaxy growth. The principal difference is in the source of carbon for GR. In the case of epitaxy growth these are carbon atoms that remain on the SiC surface after silicon atoms escape in sublimation. In case of the CVD growth, carbon comes from decomposition of methane molecules on the SiC surface. Both methods have a drawback, which consists in the need of high growth temperature. At using our CVD apparatus the necessary growth temperature for quality growth was evidently not reached. Nevertheless, we succeeded in preparing a FLG film, which consists of 5 carbon monolayers with relatively high failure rate. Using the XPS analysis it was proved that the layer contains minimum of oxygen and thus it is not a GR oxide.

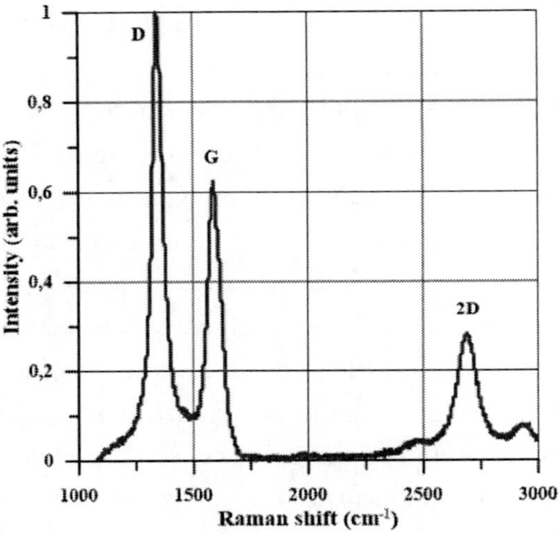

Figure 12. Raman spectrum of prepared GR film in atmosphere of pure methane.

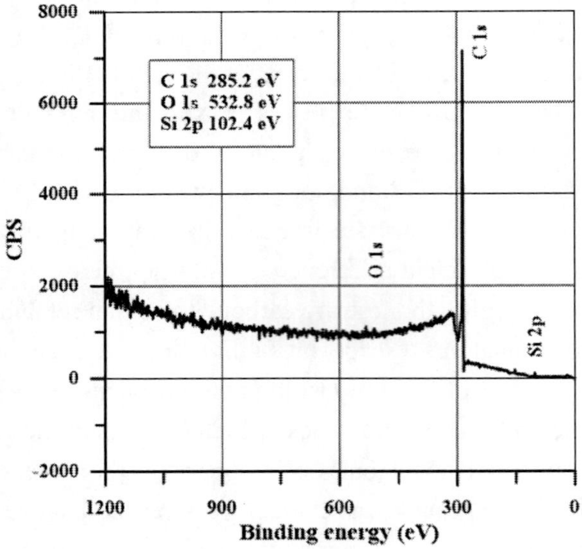

Figure 13. Survey XPS spectrum measured at perpendicular electron detection.

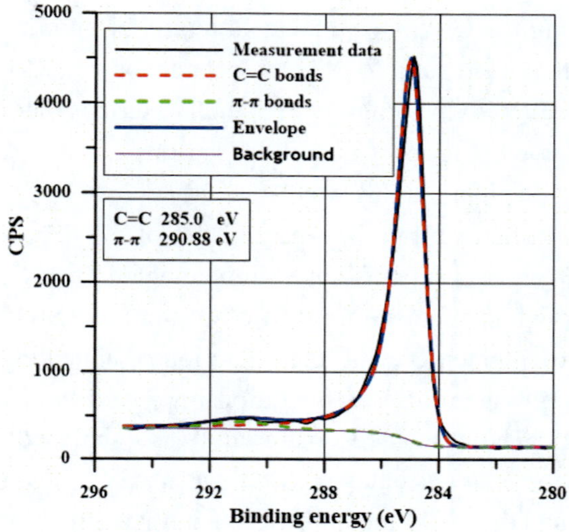

Figure 14. XPS C 1s core level spectrum.

CONCLUSION

The submitted text briefly discusses the material of a near future – GR and its preparation, based mainly on the CVD method. The introduction shortly outlines the basic properties of GR. Further, the work describes the most important methods of preparation or growth of GR. Particular attention is paid to the CVD method, on which this work is focused. The CVD method is currently being used in commercial preparation of GR films for various applications [18, 29]. The presented work describes growth of GR and its properties in the above-cited nanoCVD-8G apparatus (Moorfield Nanotechnology Ltd) with a cold-wall reactor.

Paragraph 4 describes a possible method of transfer of GR prepared by the CVD growth from copper foil onto the SiO_2/Si substrate. This transfer is necessary, as the result of the CVD growth itself is GR on a metal (most frequently on a Cu foil). From this point of view, the operation of GR transfer is the key one of its preparation. Subsequently, paragraph 5 describes a way, how it is possible to avoid the transfer using so-called transfer-free method. This way we managed to prepare BLG with good parameters using the Co/SiO_2/Si basic structure. A number of experiments further prove that it is useful to treat the copper foil prior to the GR growth. This issue is dealt with in paragraph 6, which describes the positive effect of electrochemical etching of Cu foil in various chemicals on the basic parameters of GR prepared this way.

Paragraph 7 then summarizes parameters of GR film prepared by the classical CVD process, in this case supplemented with Cu foil oxidation prior to the growth itself. After transfer of GR onto the SiO_2/Si substrate the process was complemented with the operation of annealing at temperature of 500°C in high vacuum. This operation improved the parameters of the prepared GR. The obtained material was analysed using Raman spectroscopy; the Van der Pauw method was used for electrical parameters measurement and the AFM method was used for verification of the positive effect of high-temperature annealing onto the GR structure. In the end of this paragraph, the band structure of GR is presented using the k-PEEM method.

The last paragraph of the work shows the possibility of utilization of the CVD growth method for GR preparation on SiC substrate. We managed to prepare GR even on this substrate, however with not so good parameters. Better results could be achieved at higher growth temperature, which is however out of possibilities of our apparatus. There would be certain alternative in using the PEVCD method, but we do not have it available either.

The text describes the preparation of GR by the CVD method on Cu foil. It presents the results published by the author in a number of cited publications. It is therefore an overview work, which may serve as a study material for those, who are beginning to be acquainted with GR, its properties and preparation by means of the CVD method, which is frequently being used in practice.

REFERENCES

[1] Zhan, J., et al. (2016). Carbon science in 2016: Status, challenges and perspectives. *Carbon*, *98*, 708-732.

[2] Li, X., et al. (2017). Graphene and related two-dimensional materials: Structure-property relationships for electronics and optoelectronics. *Applied Physics Reviews*, *4*, 021306.

[3] de Heer, W. A., Berger, C. & First, P. N. (2006). Patterned thin film graphite devices and method for making same. U.S. Patent US7015142B2, Mar. 21,

[4] Novoselov, K. S., (2010). Graphene: material in flat world. Nobel lecture. *The Nobel Foundation*, *8.11*.

[5] Novoselov, K. S., et al. (2004). Electric field Effect in atomically thin carbon films. *Science*, *306*(5696), 666-669.

[6] Wallace, P. R. (1947). The band theory of graphite. *Physical Review*, *71*(9), 622-634.

[7] Bolotin, K. I., et al. (2008). Ultrahigh electron mobility in suspended graphene. *Solid State Communications*, *146*(9), 351-355.

[8] Ghosh, S., et al. (2008). Extremely high thermal conductivity of graphene: Prospects for thermal management applications in nanoelectronic circuits. *Applied Physics Letters*, *92*(15), 151911.

[9] Bonaccorso, F., et al. (2010). Graphene photonics and optoelectronics. *Nature Photonics*, *4*, 611-623.

[10] Lee, C., et al. (2008). Measurement of the elastic properties and intrinsic strength of monolayer graphene. *Science*, *321*(5887), 385-388.

[11] Zheng, Q., Han, B., Cui, X., Yu, X. & Ou, J. (2017). Graphene-engineered cementitious composites: Small makes a big impact. *Nanomaterials and Nanotechnology*, *7*, 1-18.

[12] Akbar, F., et al. (2015). Graphene synthesis, characterization and its applications in nanophotonics, nanoelectronics, and nanosensing. *J Mater Sci: Mater Electron*, *26*, 4347–4379.

[13] Celis, A., et al. (2016). Graphene nanoribbons: fabrication, properties and devices. *J. Phys. D: Appl. Phys.*, *49*, 143001.

[14]Maharubin, S., et al. (2016). Synthesis and applications of semiconducting graphene. *Journal of Nanomaterials*, ID 6375962.

[15] Chen, W., Rakhi, R. B. & Alshareef, H. N. (2013). Capacitance enhancement of polyaniline coated curved-graphene supercapacitors in a redox-active electrolyte. *Nanoscale*, *5*, 4134-4139.

[16] Schniepp, H. C., et al. (2006). Functionalized single graphene sheets derived from splitting graphite oxide. *J. Phys. Chem. B*, *110* (17), 8535– 8539.

[17] Jiao, L., et al. (2010) Facile synthesis of high-quality graphene nanoribbons. *Nature Nanotechnology*, *5* (5), 321-325.

[18] Chaitanya, A., Kanishka, De Silva, K. H. & Yoshimura, M. (2020). Adlayer-free large-area single-crystal CVD graphene growth on copper. *J Mater Sci: Mater Electron*, *31*, 21821–21831.

[19] Machac, P., Fidler, T., Cichon, S. & Miskova, L. (2012). Synthesis of graphene on SiC substrate via Ni-silicidation reactions. *Thin Solid Films*, *520*, 5215-5218.

[20] Machac, P. & Pajtai, J. (2018). Graphene growth from the metal/carbon/SiO_2 structure. *J. Electrical Engineering*, *69*, 239-244.

[21] Machac, P. (2016). Graphene on metallic foils. *Nanomaterials and Nanotechnology*, *6*, 1-4.

[22] Wu, T., et al. (2016). Fast growth of inch-sized single-crystalline graphene from a controlled single nucleus on Cu–Ni alloys. *Nat. Mater.*, *15*, 43-47.

[23] Wood, J. D., et al. (2011). Effects of polycrystalline Cu substrate on graphene growth by chemical vapor deposition. *Nano Lett.*, *11*, 4547-4554.

[24] Lin, L., et al. (2016). Rapid growth of large single-crystalline graphene via second passivation and multistage carbon supply. *Adv. Mater.*, *28*, 4671-4677.

[25] Luo, B., Gao, D., Wang, H., Xu, Z. & Yu, G. (2017). Etching-controlled growth of graphene by chemical vapor deposition. *Chem. Mater.*, *29*, 1022-1027.

[26] Wu, W. (2017). Inorganic nanomaterials for printed electronics: a review. *Nanoscale*, *9*, 7342-7372.

[27] Vlassiouk, I., et al. (2013). Graphene nucleation density on copper: Fundamental role of background pressure. *J. Phys. Chem. C*, *117*, 18919-18926.

[28] Fan, L., et al. (2011). Controllable growth of shaped graphene domains by atmospheric pressure chemical vapour deposition. *Nanoscale*, *3*, 4946-4950.

[29] Ferrari, A. C., et al. (2014). Science and technology roadmap for graphene, related two-dimensional crystals and hybrid systems. *Nanoscale*, *7*, 4598-4810.

[30] Akbar, F., et al. (2015). Graphene synthesis, characterization and its applications in nanophotonics, nanoelectronics, and nanosensing. *J Mater Sci: Mater Electron*, *26*, 4347-4379.

[31] Regan, W., et al. (2010). A direct transfer of layer-area graphene. *Appl. Phys. Lett.*, *96*(11), 113102.

[32] Machac, P. & Nahlik, J. (2019). Transfer of graphene from copper foils to dielectric substrates. *Book of abstract*, *5th IMAPS Flash Conference*, *Brno University of Technology. Czech Republic*, 70-71.

[33] Kang, J., Shin, D., Baea, S. & Hong, B. H. (2012). Graphene transfer: key for applications. *Nanoscale*, *4*, 5527-5537.

[34] Machac, P., Hejna, O. & Slepicka, P. (2017). Graphene growth by transfer-free chemical vapour deposition on a cobalt layer. *J. Electrical Engineering*, *68*, 79–82.

[35] Pimenta, M., et al. (2007) Studying disorder in graphite-based systems by Raman spectroscopy. *Phys. Chem. Chem. Phys.*, *9*, 1276-1290.

[36] Gnanaprakasa, T. J., et al. (2015). The role of copper pretreatment on the morphology of graphene grown by chemical vapor deposition. *Microelectronic Engineering*, *5*, 1-7.

[37] Kwona, G. D., et al. (2014). Controlled electropolishing of copper foils at elevated temperature. *Applied Surface Science*, *307*, 731-735.

[38] Machac, P. & Tichy, T. (2018). Electochemical etching of copper foils for graphene preparation by CVD. *Proceedings of the 12th International Conference on Advanced Semiconductor Devices and Microsystems ASDAM 2018, IEEE Catalog Number CFP 18469-PRT Slovakia*, 21-24.

[39] Bointon, T. H., Barnes, M. D., Russo, S. & Craciun, M. F. (2015). High quality monolayer graphene synthesized by resistive heating cold wall chemical vapor deposition. *Adv. Mater.*, *27*, 4200–4206.

[40] Machac, P., Cichon, S., Lapcak, L. & Fekete, L. (2020). Graphene prepared by chemical vapour deposition process. *Greaphene Technology*, *5*, 9-17.

[41] Ding, D., Solis-Fernandez, P., Yanus, R. M., Hibino, H. & Ago, H. (2017). Behaviour and role of superficial oxygen in Cu for the growth of large single-crystalline graphene. *Appl. Surf. Sci.*, *408*, 142-149.

[42] Koon, D. W., Hermanova, M. & Nahlik, J. (2015). Electrical conductance sensitivity functions for square and circular cloverleaf van der Pauw geometries. *Meas. Sci. Technol.*, *26*, 115004.

[43] Machac, P. & Hrebicek, T. (2016). Synthesis of graphene on Ni/SiC structure. *J. Electrical Engineering*, *67*, 147-149.

[44] Guermoune, A., et al. (2011). Chemical vapor deposition synthesis of graphene on copper with methanol, ethanol, and propanol precursors. *Carbon*, *49*, 4204-4210.

[45] Kim, J. G., Kim, W. S., Kim, Y. H., Lim, C. H. & Choi, D. J. (2013). Formation of graphene on SiC by chemical vapour deposition with liquid sources. *Surface & Coating Technology*, *231*, 189-192.

[46] Hagihara, Y., Kajiwara, T., Visikovskiy, A. & Tanaka, S. (2013). Graphene nanoribbons grown on epitaxial $Si_xC_yO_z$ layer on vicinal SiC(0001) surfaces by chemical vapour deposition. *Appl. Phys. Express*, *6*, 055102.

[47] Liu, Q., et al. (2018). Chemical vapour deposition graphene of gradient growth method on an 4H-SiC(0001) substrate. *Appl. Surf. Sci.*, *454*, 68-73.

[48] Machac, P. (2019). Graphene prepared on SiC by chemical vapor deposition process at low temperature. *J. Electrical Eng.*, *70*, 329-331.

[49] Yang, D., et al. (2009). Chemical analysis of graphene oxide films after heat and chemical treatments by X-ray photoelectron and Micro-Raman spectroscopy. *Carbon*, *47*, 145-152.

BIOGRAPHICAL SKETCH

Petr Machac, PhD

Affiliation: University of Chemistry and Technology, Prague, Czech Republic.

Education: Czech Technical University, Prague, Czech Republic: 1977 – MSc, 1982 – PhD, 1992 – Associate Professor

Business Address: University of Chemistry and Technology, Technicka 5, Prague 6, 166 28, Czech Republic

Research and Professional Experience: 57 papers in research international journals, 2 chapters in research books.

Professional Appointments: Czech Technical University, Prague, Czech Republic 1977 – 1984.

University of Chemistry and Technology, Prague, Czech Republic 1984 – up today.

Publications from the Last 3 Years:

Machac, P; Hejna, O. Graphene growth by transfer-free CVD method using cobalt/nickel catalyst layer. *Material Science Forum*, V 919, (2018), 207-214. ISSN 0255-5476.

Machac, P; Pajtai, J. Graphene growth from the metal/carbon/SiO_2 structure. *J. Electrical Engineering*, 69, (2018), 239-244.

Machac, P. Graphene prepared on SiC by chemical vapor deposition process at low temperature. *J Electrical Engineering*, 70, (2019), NO4, 329-331.

Machac, P; Cichon, S; Lapcak, L; Fekete, L. Graphene prepared by chemical vapour deposition process. *Graphene Technology*, 5, (2020), 9-17.

Vanek, F; Jati, GNP; Okada, K; Xie, Y; Zhu, W; Machac, P; Marin, E; Pezzoti, G. Transfer-free layered graphene on silica via segregation through a nickel film for electronic application. *ACS Applied Nano Materials,* 3 (2020), 9984-9992.

Svanda, J; Kalachyova, Y; Ajami, A; Hulinsky, W; Machac, P; Siegel, J; Kolska, Z; Slepička, P; Svorcik, V; Lyutakov, O. Nonlinear optical properties and markedly higher two photon absorption of ordered c-shaped plasmon-active metal nanostructures. *Optical Materials*, 112, (2021), 110695.

INDEX

A

B

C

D

E

F

G

H

I

T

U

V

W

X

Y